MANTLE AND LOWER CRUST EXPOSED IN OCEANIC RIDGES AND IN OPHIOLITES

Petrology and Structural Geology

VOLUME 6

Series Editor:

ADOLPHE NICOLAS

Department of Earth and Space Sciences,
University of Montpellier, France

The titles published in this series are listed at the end of this volume.

Mantle and Lower Crust Exposed in Oceanic Ridges and in Ophiolites

Contributions to a Specialized Symposium of the VII EUG Meeting, Strasbourg, Spring 1993

Edited by

R.L.M. VISSERS

Faculty of Earth Sciences,
Geodynamics Research Institute,
Utrecht, The Netherlands

and

A. NICOLAS

Laboratoire de Tectonophysique,
Université des Sciences et Techniques du Languedoc,
Montpellier, France

Springer-Science+Business Media, B.V.

A C.I.P. Catalogue record for this book is available from the Library of Congress

ISBN 978-90-481-4557-7 ISBN 978-94-015-8585-9 (eBook)
DOI 10.1007/978-94-015-8585-9

Published by Kluwer Academic Publishers,
P.O. Box 17, 3300 AA Dordrecht, The Netherlands.

Kluwer Academic Publishers incorporates
the publishing programmes of
D. Reidel, Martinus Nijhoff, Dr W. Junk and MTP Press.

Sold and distributed in the U.S.A. and Canada
by Kluwer Academic Publishers,
101 Philip Drive, Norwell, MA 02061, U.S.A.

In all other countries, sold and distributed
by Kluwer Academic Publishers Group,
P.O. Box 322, 3300 AH Dordrecht, The Netherlands.

Printed on acid-free paper

Table of Contents

Introduction

This volume follows a Specialized Symposium on "Mantle denudation in slow spreading ridges and in ophiolites", held at the XII EUG Meeting in Strasbourg, spring 1993. During the meeting it was felt that the contributions to the Symposium justified a volume presenting its main scientific achievements. The present title of the volume shows that the center of interest has slightly shifted with respect to the initial objective: in order to understand the processes involved in accretion taking place at oceanic ridges, it is crucial to study the interaction between uppermost mantle and lower crust. The approach favored here is that of petrological and structural analysis of oceanic rocks in present-day oceanic ridges combined with similar studies in ophiolites. Rock specimen collected by submersibles or dredge hauls in oceanic ridge environments provide a "ground truth". However, except for areas such as the MARK (Mid-Atlantic Ridge near Kane fracture zone) where, thanks to multiple submersible dives, the local geology is known with a precision even better than in many onshore ophiolites, mutual relationships between uppermost mantle and lower crust are poorly known. In contrast, onshore ophiolites provide a necessary large-scale picture built up over many years of structural and petrological mapping.

Systematic observation of mantle rocks near the Mid Atlantic Ridge (MAR) has yielded a surprise, as it was predicted to find an "ophiolite-type" of oceanic crust in these regions. By the late 80's it became understood that mantle denudation and exposure of mantle rocks at the ridge axis result from cyclic magmatic activity. During a magmatic stage, expansion is accommodated by injection of fresh basaltic material creating the "ophiolite-type" oceanic crust, whereas during a following amagmatic stage expansion relies on tectonic stretching. Oceanic crust is then detached along movement zones leading to exhumation and exposure of the underlying mantle. In their respective papers, M. Cannat and J.F. Casey, P. Tartarotti and C. Mevel take advantage of the exposure of deep-seated rocks along the axis of the Mid Atlantic Ridge to describe the magmatic and tectonic processes which affected these deep zones of the ridge. Obviously, magmatic and tectonic processes go hand in hand at fast spreading ridges and, at least on the large scale, there is no manifestly cyclic activity. Exposure of deep rocks in such cases is much more limited as in the Deeps of the Easter Microplate where they are studied in this volume by M. Constantin and his co-workers. These three marine studies of present-day oceanic environments are followed in the volume by three onshore ophiolite studies, the first two by R. Caby and G. Molli dealing with western Alps-Apennines ophiolites probably akin to slow-spreading ridges and therefore comparing best with the MAR, and the last one by B. Ildefonse and his co-workers focussed on part of the Oman ophiolite, probably a predecessor of the fast-spreading East Pacific Rise. Eventually, the volume closes with a numerical model study of cyclic oceanic activity, developed by C. Tisseau and T. Tonnerre, involving a non steady-state thermal model for the axial domain of spreading ridges in which accretion is simulated as the superposition of seafloor spreading and thermal inputs which vary through time following magmato-tectonic cycles. Their model effectively reconciles the extreme cases of slow and fast spreading ridges.

Essentially based on the analysis of the "real rocks", this volume dealing with mantle-crust interactions reinforces the well-known contrast between processes taking place at slow and fast oceanic ridges.

We are greatly indebted to Paul van Oudenallen, Fred Trappenburg, Izaak Santoe and Brigitte Benders of the Audiovisual Service, Institute of Earth Sciences, Utrecht, for their high-quality technical support in the final stages of the preparation of this volume.

R.L.M. Vissers and A. Nicolas

Part I

Marine Studies

An Ultramafic Lift at the Mid-Atlantic Ridge: Successive Stages of Magmatism in Serpentinized Peridotites from the 15°N Region.

MATHILDE CANNAT AND JOHN F. CASEY*

URA 736, Laboratoire de Pétrologie, UPMC, 4 pl. Jussieu, 75252 Paris Cedex 05, France

**Department of Geosciences, University of Houston, Houston, TX 77204-5503, USA*

Abstract

We use the mineralogy, textures and chemical diversity of a set of gabbroic to trondhjemitic samples intrusive into serpentinized ultramafics at 15°N along the Mid-Atlantic Ridge to constrain a model for the building of the lower crust at magma-starved ocean ridges. This model involves successive stages of magmatism within mantle rocks that are rising up from the asthenosphere, and throughout the axial lithosphere, to ultimately form seafloor exposures. The relationships between gabbro chemistry and the degree of mantle melting, and the possible role of mechanical melt segregation to produce evolved magmas in this region of the Atlantic are also discussed.

Introduction

The number of documented exposures of serpentinized peridotites along the Mid-Atlantic Ridge has steadily increased in the last years, as more dredging and submersible cruise were carried on. The current interpretation is that emplacement of these rocks in the seafloor characterizes ridge regions suffering, or having suffered, a deficit in magma (Karson et al., 1987; Dick, 1989; Karson, 1991). Submersible surveys, sample studies and field data from ophiolitic massifs (Lagabrielle and Cannat, 1990; Boudier et al., 1989) suggest that, instead of being made of successive layers covering the mantle as proposed by participants to the 1972 Penrose Conference (Penrose, 1972), the crust formed at these magma-starved mid-ocean ridges may be made of discrete intrusions within mantle peridotites (Dick et al., 1989; Lagabrielle and Cannat, 1990; Mével et al., 1991; Cannat et al., 1992; Cannat, 1993). Studies of a relatively sparse sample set has shown that these intrusions display a wide range of compositions, suggesting extensive closed-system fractionation and some degree of chemical interaction with their host ultramafics (Cannat et al., 1992; Tartarotti et al., this volume). The investigation of the mineralogy, chemical diversity and modes of emplacement of

R.L.M. Vissers and A. Nicolas (Eds.),
Mantle and Lower Crust Exposed in Oceanic Ridges and in Ophiolites, 5-34.

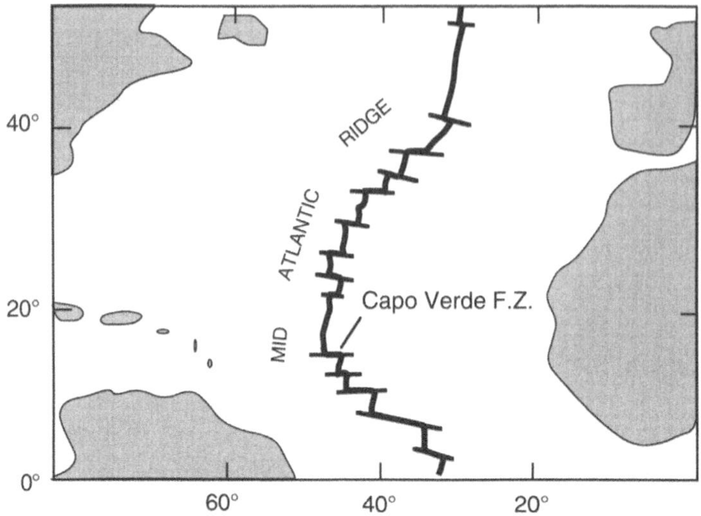

Figure 1. General map of the Atlantic with location of Capo Verde Fracture Zone.

these gabbroic to dioritic and trondhjemitic rocks is of primary interest, as they likely are a significant component of the lower magmatic plumbing system at magma-poor oceanic ridges, below the upper crustal extrusive and effusive basalts.

The sample set collected during the FARANAUT Nautile cruise in the Capo Verde Fracture zone region (15°N; Fig. 1) of the Mid-Atlantic Ridge (Bougault et al., 1993) is the most complete available to date. It comprises 127 samples of ultramafics, gabbroic rocks and trondhjemites, many of which preserve intrusive relationships between the ultramafics and gabbroic or trondhjemitic dikes. The 15°N sample set also includes rare diabase and basalt dikes, cutting the ultramafics and presumably acting as feeder dikes for axial valley floor basalt flows. The chemistry of these dikes, and their possible petrogenetic links with coarser grained intrusions, will be detailed in a forthcoming paper (Casey et al., in prep.). The sampled ultramafic suite also includes occasional wehrlites similar to rocks found in other oceanic locations (Girardeau and Francheteau, 1993), and at the crust-mantle transition zone in ophiolites (Casey et al., 1981; Nicolas and Prinzhofer, 1983; Benn et al., 1988). We use the primary and secondary mineralogy of these different types of magmatic rocks, the nature of chemical interactions with their host peridotite, and the timing of their injection relative to deformational and metamorphic events, to discuss the modes of emplacement of crustal rocks in mantle-derived ultramafics from the 15°N region. Comparison with observations made on a smaller variety of samples in the MARK (Mid-Atlantic Ridge/Kane Fracture Zone) area (Mével et al., 1991; Tartarotti et al., this volume) suggests that the picture we draw in the discussion of this paper may be generally applicable to magma-poor portions of the Mid-Atlantic Ridge.

Gabbroic rocks from the Capo Verde Fracture Zone region (Fig. 1) locally form large bodies (more than 100 m wide) of coarse to medium grained rocks, surrounded by serpentinized peridotites. Other smaller (decimeter to meter-sized) gabbroic and trondhjemitic intrusives, and veins (5 cm thick or less) are observed at the scale of outcrops, or of individual samples, respectively. In general, intrusive contact relationships are well documented for the smaller intrusives, but not for the larger bodies. Based on the mapped distribution of lithologies (Cannat et al., unpubl. data), however, it appears likely that large gabbroic bodies are also intrusive into serpentinized peridotites. Mineralogically, intrusions over 10 cm in width range from gabbroic to trondhjemitic (Table 1). Smaller dikes and veins are extensively altered and their plutonic primary mineralogy can only be inferred from preserved accessory minerals, or from the composition of hydrothermal replacement minerals. This explains why a good place is made to the secondary mineralogy of these veins in the descriptive part of this paper.

Geological Setting

South of the Capo Verde fracture zone

Serpentinized peridotites crop out on both walls of the nodal basin south of the Capo Verde fracture zone (Fig. 2). Serpentinized peridotites also form extensive outcrops on a series of topographic highs which form the southern transform valley wall (Bougault et al., 1993; Dick et al., unpublished report). Samples studied here were collected along the tracks of Nautile dive FR8 on the western wall of the nodal basin, and of Nautile dives FR1 to FR9 on the flank of the easternmost transform wall high (Fig. 2). Pillows basalts are exposed at the foot (dive FR7) and on top of this high (dive FR1). Gabbroic rocks do not form massive outcrops, but decimeter to millimeter-sized dikes and veins in the ultramafics (Table 1). These ultramafics are mostly serpentinized pyroxene-poor harzburgites, but serpentinized dunites and wehrlites also occur (Table 2).

North of the Capo Verde fracture zone

Serpentinized peridotites crop out extensively along the western wall of the Mid-Atlantic Ridge axial valley north of the Capo Verde Fracture Zone (Fig. 3). Ultramafic outcrops are also documented around 15°37'N on the eastern rift valley wall (dives FR10, FR11 and FR23; Fig. 3), and on a ridge-parallel cliff located a few kilometers off-axis (dive FR12) and interpreted as a fossil median valley wall. These ultramafic outcrops comprise mostly serpentinized pyroxene-poor harzburgites, with only occasional serpentinized dunites (Table 2), The proportion of plutonic (dominantly gabbroic) intrusives is far greater than in the region south of the Capo Verde fracture zone

Table 1. Brief description of gabbroic rocks, trondhjemite and veins sampled along the Nautile dives.

sample #	*depth (m)*	*primary and (alteration) mineralogy*	*accessory minerals*	*ductile deformation (recrystallized minerals)*
Gabbroic rocks in massive outcrops				
FR12-1	4003	gabbronorite		weak (CPX,PL)
FR12-2	3860	gabbronorite	IL	strong (CPX,OPX,PL)
FR12-3	3719	gabbronorite	AP, IL	weak (PL)
FR12-5	3512	gabbro		no
FR12-6	3500	gabbronorite		weak (PL)
FR12-7	3460	gabbronorite	IL (QZ, BI, ZR)	strong (CPX,OPX,PL,AM)
FR12-8	3413	gabbronorite		weak (PL)
FR12-9	3426	gabbronorite	ZR	strong (CPX,OPX,PL)
FR12-10	3322	gabbronorite	SULF	weak (PL)
FR16-2	4342	gabbronorite	AM (QZ, BI, S)	strong (CPX,OPX,PL)
FR16-3	4070	gabbronorite	(QZ, BI, ZR, S)	strong (PL)
Dikes and meter-sized intrusions in ultramafic outcrops				
FR3-4A	3200	gabbro		no
FR7-10	3600	gabbro (AM,EP,CH)	ZR	strong (AM)
FR9-6	3804	gabbro (AM,EP,CH)	AP, IL	no
FR16-4B	3629	gabbro (AM,PL)	IL,AP,AM,S	strong (CPX,AP,IL,AM,PL)
FR16-5	3596	trondhjemite (AM,CH)	ZR	no
FR16-6	3588	trondhjemite	FK,IL	no
FR22-6	3705	gabbronorite (AM,CH,EP)	AP, ZR, AM, IL	strong (CPX,OPX,PL,AM,AP,IL)
Veins in ultramafic outcrops				
with pseudomorphs afer pyroxene:				
FR7-15	3373	? (AM,EP,CH)	ZR	no
FR8-2	4625	? (AM,CH)	PH, IL	no
FR8-13	3428	? (AM,CH)	ZR, AM	strong (AM,PL,PX)?
FR9-7	3620	? (AM,CH,EP,SPH,CPX)		strong (AM,CH,SPH)
FR10-7	3624	? (AM,CH)	AP, IL, AM	strong (AM,PL,PX)?
FR10-9	3500	? (AM,CH)	ZR	no
FR21-6	3910	? (AM,CH)	AP, AM, IL	strong (AM,PL,PX)?
FR22-4	3671	? (AM,CH)	ZR, AP, IL	no
FR23-2	3897	? (AM,CH)	ZR	weak (AM)
with abundant (primary ?) brown amphibole:				
FR21-14	3354	amphibolite (AM, CH)	IL	no
with abundant (primary?) euhedral greenish-brown amphibole:				
FR16-4C	3629	? relic AM (AM,CH)	ZR, IL	no
FR17-4	3708	? relic AM (AM,CH)	ZR, S	no
with no relics or pseudomorphs of primary minerals:				
FR10-4	3775	? (AM)		weak (AM)
FR16-4A	3629	? (AM, CH)	PH, ZR, S	weak (AM,CH)
FR19-1	3391	? (AM,CH)		no
FR23-4	3703	? (CH)		no
with abundant euhedral colorless amphibole:				
FR5-4	3903	AM+OPX		no
FR9-4	3744	AM		no
FR19-4	3436	AM		no
FR23-5	3640	AM	IL	no

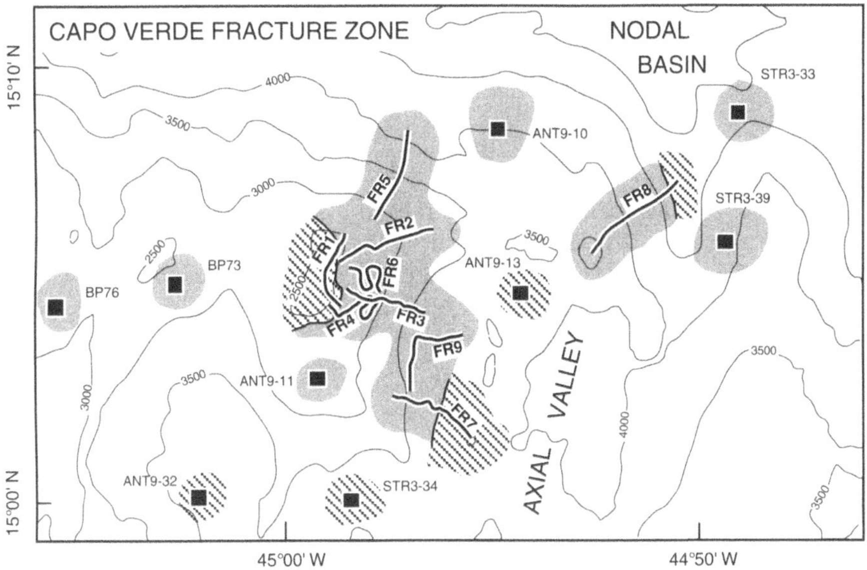

Figure 2. Simplified bathymetric map (Bougault et al., unpubl. data) of the Mid-Atlantic Ridge south of the Capo Verde Fracture Zone, contoured at 500 m intervals, showing locations of Faranaut Nautile dives FR1 to FR9. Also shown are locations of Russian dredges (BP73 and 76, STR3-33, 34 and 39, and ANT9-10, 11, 13 and 32) reported by Dick et al.(unpubl. report, 1991). A schematic geological interpretation is proposed, with ultramafic and gabbroic outcrops in gray, and basalt flows in hatchures.

(Table 1). Massive gabbroic outcrops are documented by Nautile dives FR12 on the fossil eastern median valley wall, and FR16 on the active western wall (Fig. 3 and Table 1). During these two dives, serpentinized peridotites were

Table 2. Serpentinized dunites and wehrlites sampled along the Nautile dives

Sample #	*depth (m)*	*Sample #*	*depth (m)*	*Sample #*	*depth (m)*
Serpentinized Dunites					
FR1-5	2482	FR4-3	2961	FR9-3	3756
FR1-7	2454	FR6-8	2709	FR9-4	3744
FR2-4	3052	FR6-10	2690	FR9-5	3804
FR3-3	3411	FR6-11	2595	FR9-7	3620
FR3-4B	3200	FR7-12	3561	FR9-8	3596
FR3-5	3011	FR8-8	3974	FR21-17	3265
FR3-7	2731	FR9-1	3991	FR22-5	3710
FR4-1	2906	FR9-2	3943		
Serpentinized Wehrlites					
FR7-15	3373	FR9-6	3804		

sampled above the gabbroic outcrops, or intercalated with gabbroic outcrops that may therefore be interpreted either as tectonic intercalations along median-valley wall normal faults, or as discrete intrusions within the ultramafics (our preferred interpretation). Gabbroic rocks and trondhjemites also occur as decimeter to meter-sized dikes and intrusions in the ultramafics explored during dives FR16 and FR22 on the western median valley wall (Table 1). Such decimeter to meter-sized gabbroic intrusions were also dredged on top of the eastern axial valley wall around 15°37'N (dredge RD88-Dr8; Fig. 3; Cannat et al., 1992). Finally, veins of possible plutonic origin (Table 1) have been identified in about 40% of the ultramafic samples collected in this northern region.

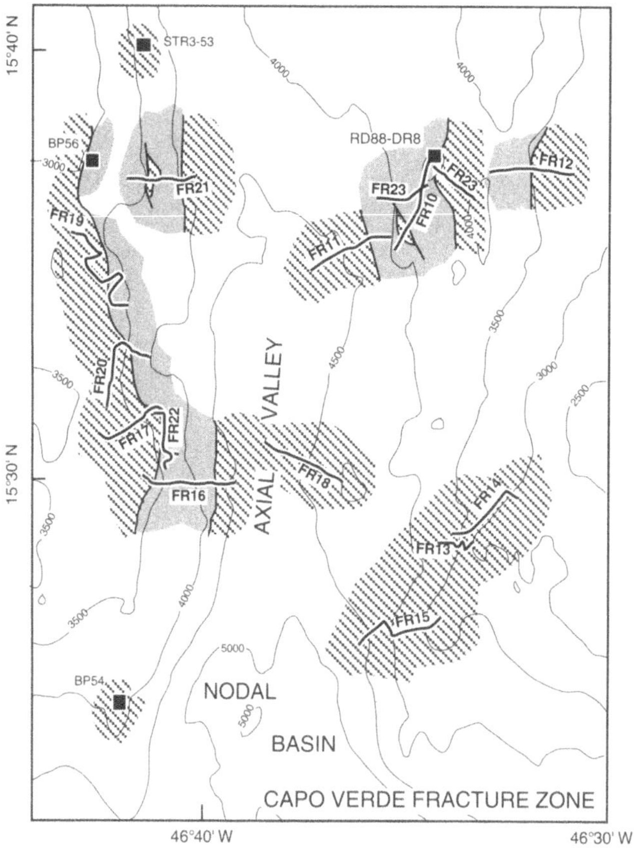

Figure 3. Simplified bathymetric map (Bougault et al., unpubl. data) of the Mid-Atlantic Ridge north of the Capo Verde Fracture Zone, contoured at 500 m intervals, showing locations of Faranaut Nautile dives FR10 to FR23. Also shown are locations of Russian dredges (BP54 and 56 and STR3-53) reported by Dick et al.(unpubl. report, 1991), and of the R.V. Jean Charcot dredge RD88-Dr8. A schematic geological interpretation is proposed, with ultramafic and gabbroic outcrops in gray, and basalt flows in hatchures.

Texture and Mineralogy of Ultramafic Rocks

Serpentinized harzburgites, dunites and wehrlites collected south of the Capo Verde Fracture Zone (Fig. 2) have coarse protogranular to porphyroclastic textures, with millimeter to centimeter-sized olivine and pyroxene grains. These ultramafics are partially altered to a mixture of serpentine and oxidized clays, with crosscutting veins of carbonate minerals. Alteration varies from pervasive to moderate (30-40%). Serpentinized harzburgites have a poorly defined spinel and orthopyroxene foliation. Olivine forms centimeter-sized irregular grains, and smaller polygonal grains (1 to 3 mm in size) with frequent 120° triple junctions. Subgrain boundaries in olivine are generally scarce and widely spaced. The crystallographic fabric of olivine is commonly strong. Orthopyroxene grains are millimeter to centimeter-sized and lobate in shape, with cuspate or straight grain boundaries with the surrounding olivine. Clinopyroxene is generally absent, occurring only in a few samples as rare exsolution lamellae in orthopyroxene, or as small interstitial grains associated with orthopyroxene and spinel. Spinel is lobate and commonly associated with orthopyroxene, suggesting that it originated from the breakdown of garnet in the spinel stability field (Kushiro and Yoder, 1966). Such textures, also common in ophiolite massifs, are inferred to have formed in the asthenospheric mantle (Mercier and Nicolas, 1975).

Serpentinized dunites collected south of the Capo Verde Fracture Zone commonly display a weak spinel foliation. Olivine grains are large (up to 2 cm), with rare subgrain boudaries. Their crystallographic fabric is weak or absent. Spinel grains are rounded to euhedral. Clinopyroxene in serpentinized wehrlites forms centimeter-sized poikiloblasts (Fig. 4a). Plagioclase is interstitial and commonly displays mechanical twins. Olivine grains are millimeter to centimeter-sized, rounded in shape, with rare subgrain boundaries and no crystallographic fabric.

Serpentinized harzburgites and dunites collected north of the Capo Verde Fracture Zone (Fig. 3) are more extensively serpentinized than the ultramafics collected south of this Fracture Zone. Spinel is the only relic mineral in many samples. Texturally, these northern ultramafics are also different from the southern ones: they have a strong foliation, marked by elongated orthopyroxene porphyroclasts and spinel trains, in a matrix of olivine and orthopyroxene neoblasts less than 0.1 mm in size. Olivine and orthopyroxene crystallographic fabrics are strong. Fabrics due to this deformation event have been studied in samples from dredge RD88-Dr8 (Fig. 3), and are inferred to have developed in the deep lithosphere of the axial region (Cannat et al., 1992). Spinel is lobate and commonly associated with orthopyroxene grains. Clinopyroxene occurs as rare and thin exsolution lamellae in orthopyroxene porphyroclasts, and as equally rare neoblasts in the recrystallized matrix.

Figure 4. Microphotographs showing magmatic rocks from the 15°N region: (a) serpentinized wehrlite sample FR7-15, with zircon-bearing vein containing mesh of secondary amphibole and chlorite and of talc and chlorite near contact with host wehrlite, crossed nicols; (b) gabbronorite sample FR12-2 showing limited ductile deformation along grain boundaries of igneous plagioclase and pyroxenes, and two types of plagioclase neoblasts: relatively coarse polygonal ones (about 0.3 mm) and later smaller ones (about 0.03 mm), crossed nicols; (c) gabbronorite sample FR16-3 with ductile-brittle deformation causing recrystallization of small plagioclase neoblasts and fracturing of plagioclase porphyroclasts.

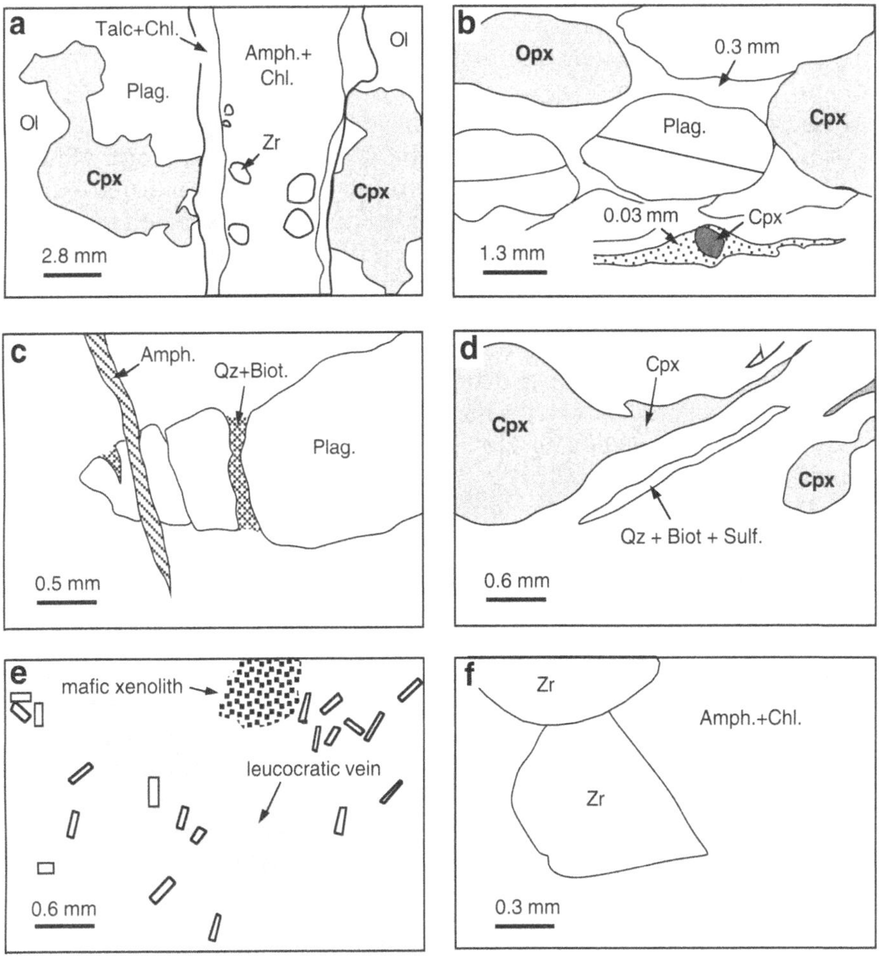

Figure 4. (continued) Fractures are filled with quartz and biotite or with green-brown amphiboles, crossed nicols; (d) gabbronorite sample FR12-7 showing ductile shear zone in fine-grained plagioclase and amphibole neoblasts with rotated clinopyroxene porphyroclasts and a thin concordant vein of quartz, biotite and sulphides, plain polarized light; (e) trondhjemite sample FR16-6, with finer-grained more mafic xenoliths and a leucocratic vein. Shape fabric of plagioclase tablets define a crude magmatic lineation, crossed nicols: (f) zircon in mesh of secondary amphibole and chlorite, in vein from serpentinized wehrlite sample FR7-15, crossed nicols.

Texture and Mineralogy of Gabbroic Rocks, Trondhjemites and Veins in the Ultramafics

Gabbroic rocks in massive outcrops

Massive outcrops of gabbroic rocks have been sampled along the tracks of dives FR12 and FR16 north of the Capo Verde transform (Fig. 3). These are medium to coarse grained (0.1 to 2 cm) gabbronorites with anhedral to euhedral pyroxenes. Elongated pyroxenes in some samples define a foliation and lineation which we interpret as magmatic in view of the lack of associated intracrystalline deformation. A diffuse grainsize layering, sub-parallel to this foliation, was observed in a few samples. Plagioclase is tablet-shaped or anhedral, and often partly recrystallized into polygonal neoblasts. Apatite, ilmenite, brown amphibole, zircon and sulphides occur as accessory minerals. Three samples also contain diffuse patches and veinlets of a quartz-biotite assemblage, with accessory zircon and sulphides.

Ductile deformation affected most samples (Table 1), causing the partial recrystallization of plagioclase and producing in some samples a foliation and lineation which obscures the original magmatic texture. Two successive synkinematic recrystallization events can be identified. During the first event, recrystallization of plagioclase in relatively large polygonal grains (0.2-0.3 mm; Fig. 4b) was accompanied in the most deformed samples by partial recrystallization of orthopyroxene and clinopyroxene (Table 1). The second recrystallization event produced much smaller plagioclase neoblasts (Fig. 4b), and microfracturation of plagioclase and pyroxene porphyroclasts. In sample FR12-7, synkinematic crystallization of amphibole also occurred during this brittle/ductile event. Limited overprinting of the earlier recrystallization event by this later deformation is clear in many samples, with two sizes of plagioclase neoblasts (Fig. 4b).

Quartz, biotite and accessory zircon and sulphides in samples FR12-7, FR16-2 and FR16-3 (Table 1) occur as diffuse patches and veinlets, some of which crosscut plastically deformed and recrystallized intervals, while others fill tension cracks in plagioclase and pyroxene porphyroclasts (Fig. 4c). Such cracks, which do not cross into adjacent recrystallized domains, formed during late stages of the brittle/ductile deformation event. One exception occurs in sample FR12-7 where ductile deformation appears to have continued after crystallization of quartz-biotite veinlets, which are transposed into shear bands made of a plagioclase-amphibole recrystallized assemblage (Fig. 4d) .

Finally, most gabbro samples are cut by microcracks filled with brownish-green amphiboles. These microcracks cut through recrystallized intervals and therefore postdated ductile deformation. Hydrothermal fluids circulating in these cracks induced limited replacement of neighbouring pyroxenes by amphibole.

Gabbroic dikes

The three gabbro dikes collected south of the Capo Verde transform (Table 1) contain clinopyroxene, anhedral or poikilitic, and plagioclase in medium to coarse grained (0.4 to 3 cm) euhedral to anhedral crystals. They have no magmatic fabric. Sample FR7-10 also contains zircon, and sample FR9-6 has abundant apatite and some ilmenite. Sample FR3-4A is relatively fresh, but samples FR7-10 and FR9-6 are extensively altered into amphibole and chlorite. Chlorite, associated with epidote, also fills veins crosscutting these two samples. Still later veins, filled with quartz and albite, cut the epidote veins in sample FR7-10. Alteration is maximal close to contacts with the host ultramafics, obscuring the textural and mineralogical nature of these contacts. The altered gabbro in sample FR7-10 has been deformed in a brittle-ductile manner, with fracturing of plagioclase, and dynamic recrystallization of secondary amphiboles. The two other samples are cut by dense arrays of microfractures, but show no ductile deformation.

Only two decimeter-sized gabbroic dikes have been collected during dives north of the Capo Verde transform. Decimeter-sized intrusions of an iron-rich, zircon and apatite-bearing olivine diorite have also been sampled in this northern region by dredge RD88-Dr8 (Fig. 3; Cannat et al., 1992). Sample FR16-4B is an altered gabbro with a strong ductile foliation. It contains centimeter-sized aggregates of anhedral millimeter-sized clinopyroxene with sulphides inclusions, in a foliated matrix of finely recrystallized plagioclase and amphibole, with disseminated ilmenite, apatite and clinopyroxene grains now extensively replaced by amphibole. Some amphibole pseudomorphs after clinopyroxene contain minute inclusions of brown amphibole that may be primary. The synkinematically recrystallized assemblage comprises small plagioclase neoblasts (0.02 mm) and brownish-green prismatic amphibole. Because the sample's foliation is also accentuated by streaks of amphibole pseudomorphs after clinopyroxene neoblasts and of fine-grained ilmenite and apatite, it is likely that ductile deformation started with partial recrystallization of the magmatic mineral assemblage. Post-kinematic alteration into acicular amphibole and albitic plagioclase is extensive. Sample FR22-6 is a foliated gabbronorite, with rounded orthopyroxene and clinopyroxene porphyroclasts in a recrystallized matrix of plagioclase, orthopyroxene, clinopyroxene and accessory ilmenite, brown amphibole and apatite. The size of plagioclase neoblasts (about 0.2 mm) is similar to that of neoblasts produced during the early ductile event in massive gabbroic outcrops. The synkinematically recrystallized assemblage is now partially altered into undeformed acicular amphibole, chlorite and occasional epidote. Sample FR22-06 also contains xenoliths of its host serpentinized harzburgite, and is cut by undeformed serpentine veins. Clean and straight grain boundaries with 120° triple junctions between finely recrystallized olivine and orthopyroxene in the harzburgite xenoliths, and pyroxene, amphibole and plagioclase neoblasts in the dike indicate that ductile deformation in this sample also affected the host peridotite.

Trondhjemite intrusions

Samples FR16-5 and FR16-6 come from meter-sized fine-grained (0.1 to 0.4 mm) trondhjemite intrusions, with 20-30% quartz. Sample FR16-5 is extensively altered into chlorite and acicular amphibole. Sample FR16-6 is fresh and contains euhedral plagioclase, biotite, and greenish brown amphibole (Fig. 4e). It is not deformed, but tablet-shaped plagioclase crystals outline a faint magmatic lineation. It contains centimeter-sized xenoliths of a more mafic and finer grained biotite and amphibole assemblage, with ilmenite inclusions in amphiboles. Sample FR16-06 also contains diffuse veins of a more acidic quartz and plagioclase assemblage (Fig. 4e), with biotite and potassic feldspar as accessory minerals.

Veins

Veins of possibly plutonic origin have been sampled in serpentinized ultramafics both south and north of the Capo Verde transform (Table 1). Contacts with the host ultramafics are lined with a mesh of talc, serpentine, acicular amphiboles and occasional chlorite. The thickness of these contact zones varies from one sample to another between less than 1 mm (Fig. 4a) to over 2 cm. The veins themselves are extensively altered into amphibole and chlorite with occasional epidote, sphene, or secondary clinopyroxene (sample FR9-7). All veins crosscut and therefore postdate the ductile deformation fabric of their ultramafic host. Veins and their altered contact zones are systematically cut by serpentine veins and tension cracks, indicating that serpentinization-induced swelling of their ultramafic host postdated the emplacement and the early alteration of the veins.

Some veins contain amphibole pseudomorphs after primary pyroxenes (Table 1) and are inferred, based on these pseudomorphs and on their accessory mineralogy, to have a gabbroic or dioritic protolith. Similar veins have been collected in dredge RD88-Dr8 (Fig. 3; Cannat et al., 1992). Other veins do not contain amphibole pseudomorphs after primary pyroxenes, but contain relics of abundant brown

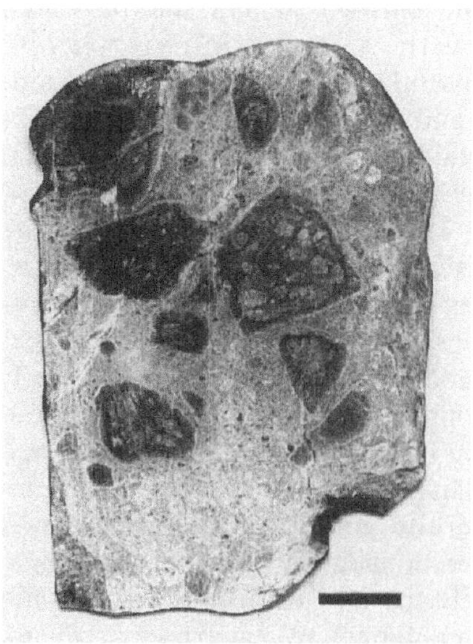

Figure 5. Photograph of sample FR17-4 showing angular to rounded fragments of serpentinized harzburgite in a fine-grained mesh of euhedral green-brown amphibole, secondary acicular amphibole, talc, chlorite and accessory zircon and sulphides. Scale bar 1 cm.

hornblende, euhedral greenish-brown amphibole, or euhedral colorless amphibole, preserved in a mesh of acicular amphibole with or without chlorite. Finally, four vein samples contain neither relics nor pseudomorphs of their pre-alteration mineral assemblage, but occasional zircon, phlogopite and sufides as accessory minerals. These different types of veins are present in ultramafics from both north and south of the Capo Verde Fracture Zone and crosscut all ultramafic lithologies (serpentinized harzburgites, dunites and wehrlites).

Veins with pseudomorphs after pyroxenes

Samples FR8-13, FR10-7 and FR21-6 have fine-grained foliated textures, with streaks of rounded chlorite and amphibole pseudomorphs after pyroxene, recrystallized grains of brown amphibole, apatite and ilmenite, in a groundmass of undeformed chlorite and acicular amphibole, which may have replaced finely recrystallized plagioclase and pyroxene. Texturally, and from what is preserved of their primary mineralogy, these altered veins appear similar to the dike of recrystallized gabbronorite of sample FR22-6.

Judging from its high content in secondary sphene, sample FR9-7 may have originally been an ilmenite gabbro. It is now pervasively rodingitised, with abundant epidote and secondary clinopyroxene in the groundmass and in veins. This calcium metasomatic event was associated with cataclastic deformation, and post-dates an earlier ductile deformation, with synkinematic crystallization of colorless amphibole and chlorite (Table 1).

Veins in samples FR7-15 (Fig. 4a), FR8-2, FR10-9, FR22-4 and FR23-2 have mm-sized pseudomorphs after pyroxenes, which apparently have not been deformed prior to their alte-ration. Zircon (Fig. 4f) and ilmenite occur as accessory minerals in three of these samples. Sample FR8-2 also contains phlogopite. Limited deformation with recrystallization of secondary acicular amphibole is observed in sample FR23-2.

Veins with abundant brown amphibole

The vein in sample FR21-14 is zoned, with a 1 cm-thick brown amphibolite layer against the contact with the serpentinized peridotite, and a core made of about 20% brown amphibole in a mesh of chlorite. Ilmenite occurs in small disseminated grains. Brown amphibole is partly altered into acicular colorless amphibole. There is no evidence for deformation.

Veins with abundant euhedral greenish-brown amphibole

Samples FR16-4C and FR17-4 contain small (0.3 to 0.5 mm) euhedral greenish-brown amphiboles, extensively altered into colorless amphibole, in a groundmass of acicular amphibole, chlorite and clays. In sam-

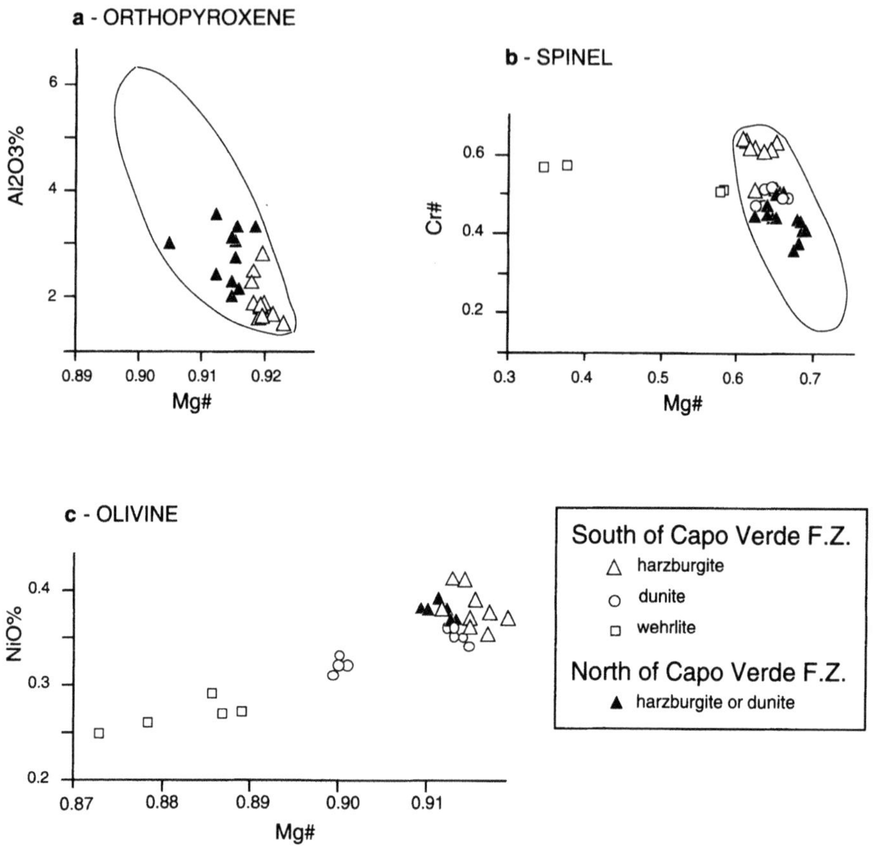

Figure 6. Mineral composition of serpentinized harzburgites, dunites and wehrlites from the 15°N region. Field of North and Central Atlantic residual peridotites (Bonatti et al., 1993; this study) is shown for orthopyroxene and spinel compositions.

ple FR17-4, this material includes angular fragments of the host serpentinized harzburgite (Fig. 5). Disseminated zircon, ilmenite or sulphides occur as accessory minerals. This fine-grained texture with small amphibole prisms is similar, except for the lack of quartz, to that of trondhjemite samples collected in the same area during dive FR16 (Fig. 3). The protolith of these veins may thus have been a fine-grained amphibole-bearing diorite. Sample FR17-4 would then be a magmatic breccia (Fig. 5). The centimeter-sized ultramafic xenoliths in this sample are surrounded by thick alteration halos with talc, serpentine, acicular amphibole and chlorite. These halos also contain relict zircon and abundant sulphides, suggesting that they developed, at least in part, from the vein material. Neither the greenish-brown amphiboles nor the alteration minerals show evidence of deformation, but amphibole prisms are locally oriented, defining a lineation that may be magmatic in origin.

Veins with abundant colorless amphibole

Veins in samples FR5-4, FR9-4, FR19-4 and FR23-5 contain euhedral and colorless amphibole, partly altered into secondary acicular amphibole. There are no accessory minerals, except for rare ilmenite grains in sample FR23-5 (Table 1). In sample FR5-4, amphibole is associated with orthopyroxene, as euhedral undeformed grains 0.5 to 1 mm in size, or as smaller grains filling cracks in the host ultramafic rock.

Veins with no relics or pseudomorphs of pre-alteration minerals

Veins in samples FR10-4 and FR19-1 contain an alteration assemblage of acicular amphibole and chlorite, with no pseudomorphs after earlier minerals. Acicular amphiboles in sample FR10-4 are kinked and partly recrystallized. Veins in samples FR16-4A and FR23-4 contain a mesh of talc, acicular amphibole, chlorite and serpentine very similar to the alteration halos observed around ultramafic xenoliths in samples FR16-4C and FR17-4 (Fig. 5). Sample FR16-4A also contains accessory zircon, phlogopite and sulphides.

Mineral Chemistry

Mineral analyses were performed on the CAMEBAX microprobe of the University of Paris VI, with an accelerating potential of 15kV and a beam current of 15 nA. Higher beam currents (40 or 80 nA) and long counting times (20 to 60 s) were used for pyroxenes, spinel and olivine in ultramafics. Data plotted in Figs. 6 to 8 correspond to selected analyses, covering the range of chemical variations observed in each sample. Tables of these selected analyses are available upon request to the authors.

Serpentinized harzburgites, dunites and wehrlites

Pyroxene
Orthopyroxene in serpentinized harzburgites collected south of the Capo Verde fracture zone has high Mg# and low Al_2O_3 contents and plots (Fig. 6a) at the most depleted end of the North and Central Atlantic residual peridotites trend (Bonatti et al., 1993; this study). Orthopyroxene in serpentinized harzburgites collected north of the Capo Verde fracture zone is less depleted in aluminium and iron (Fig. 6a). Clinopyroxene is scarce in both groups of harzburgitic samples; compared with clinopyroxene from the serpentinized wehrlites, it has low Na_2O contents (0.03 to 0.06% against 0.29 to 0.46% in the wehrlites), low TiO_2 contents (0.03 to 0.08% against 0.09 to 0.7% in the wehrlites), but similar Al_2O_3 and Cr_2O_3 contents (2.53 to 4.65% and 1.08 to 1.64% respectively). Clinopyroxene with a relatively high Na_2O

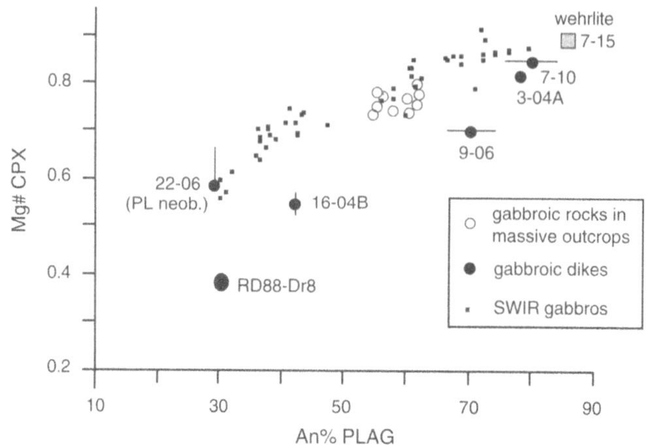

Figure 7. Clinopyroxene Mg# versus plagioclase An content in gabbroic rocks from the 15°N region and in serpentinized wehrlite sample FR7-15. SWIR (Southwest Indian Ridge) data from ODP Site 735B (Ozawa et al., 1991) and other Southwest Indian Ridge gabbroic rocks (Meyer et al., 1989). Error bars for samples FR7-10, FR9-6 and FR16-4B cover range of values measured in igneous clinopyroxene and plagioclase in these samples. Values plotted for sample FR22-6 correspond to igneous clinopyroxene porphyroclasts and to plagioclase neoblasts. Error bar shows range of Mg# in clinopyroxene neoblasts of this sample (see text).

content (0.18%) forms minute interstitial grains in one serpentinized harzburgite sample (FR6-1) from the southern outcrops. It could represent a small volume of interstitial melt trapped in this sample. Orthopyroxene in this sample also has higher Al_2O_3 contents than in other serpentinized harzburgites from the southern outcrops (2.2 to 2.7% against 1.4 to 2%).

Spinel

Spinel in serpentinized harzburgites collected south of the Capo Verde fracture zone has high Cr# and plots (Fig. 6b) at the most depleted end of the North and Central Atlantic residual peridotites trend (Bonatti et al., 1993; this study). Spinel in serpentinized harzburgites and dunites collected north of the Capo Verde fracture zone has lower Cr# and slightly higher Mg#. Intermediate values are found in spinels from southern serpentinized dunites, but associated with higher TiO_2 contents (0.11 to 0.27% against 0.04 to 0.09% in serpentinized harzburgites and in serpentinized dunites from northern outcrops). Intermediate Cr# and Mg# values are also found in spinel from the interstitial clinopyroxene-bearing southern harzburgite sample FR6-1. Spinel in serpentinized wehrlites has significantly lower Mg# (Fig. 6b), and higher TiO_2 contents (0.4 to 2.6%).

Olivine

Olivine in serpentinized harzburgites collected south of the Capo Verde fracture zone has high Mg# and NiO contents (Fig. 6c). Olivine Mg# and NiO contents are slightly lower in serpentinized harzburgites and dunites collected north of the Capo Verde fracture zone. Serpentinized dunites and wehrlites from the southern outcrops show a range of lower Mg# and NiO contents (Fig. 6c).

Plagioclase
Plagioclase in serpentinized wehrlites collected south of the Capo Verde fracture zone has An contents of 85 to 90% (Fig. 7).

Gabbroic Rocks, Trondhjemite and Veins in the Ultramafics

Orthopyroxene
Orthopyroxene occurs as an igneous phase and as neoblasts in samples from massive gabbronorite outcrops of dives FR12 and FR16, as neoblasts in the foliated metagabbronorite dike of sample FR22-6 (orthopyroxene porphyroclasts in this dike are extensively altered), and as a vein phase in association with euhedral colorless amphibole in sample FR5-4.

Orthopyroxene porphyroclasts from dives FR12 and FR16 gabbronorites have Al_2O_3 and TiO_2 contents of 1 to 1.45% and 0.1 to 0.45% respectively. Mg# range from 69.7 to 75.6% and CaO contents from 1.1 to 2.3%, with occasional grains of more calcic orthopyroxene (CaO contents of 5 to 6%). Orthopyroxene neoblasts have lower Al_2O_3 contents (0.3 to 1.2%) and Mg# (63.6 to 69.8%) and are compositionaly similar to neoblasts in dike sample FR22-6.

Vein orthopyroxene in sample FR5-4 has high Mg# (85 to 90%), low Al_2O_3 and CaO contents (0.3 to 0.7% and 0.2 to 0.3% respectively), and TiO_2 contents of 0.02 to 0.07%.

Clinopyroxene
In the gabbronorite outcrops of dives FR12 and FR16, igneous clinopyroxene has Mg# between 0.7 and 0.8 (Fig. 7), low Cr_2O_3 contents (less than 0.08%), Na_2O contents of 0.24 to 0.5%, and TiO_2 contents varying between 0.2 and 0.91%. In general, lower Na_2O and TiO_2 contents (0.24 to 0.33% and 0.2 to 0.55% respectively) are found in samples that also contain veinlets and patches of a quartz and biotite assemblage (Table 1). Al_2O_3 contents range between 1.9 and 2.5%. Clinopyroxene neoblasts in deformed intervals have generally lower Al_2O_3 (0.7 to 2.3%), but otherwise similar compositions.

Gabbro dikes of samples FR3-4A and FR7-10, collected in the ultramafics south of the Capo Verde Fracture Zone, have high clinopyroxene Mg# (0.81 to 0.88; Fig. 7), high Cr_2O_3 contents (0.09 to O.33%), relatively low TiO_2 contents (0.1 to 0.45%) and Na_2O contents of 0.25 to 0.3%. The gabbro dike in sample FR9-6, also from the southern ultramafic outcrops, has lower clinopyroxene Mg# (0.67 to 0.7; Fig. 7), lower Cr_2O_3 contents (0.02%), higher TiO_2 contents (0.55 to 0.89%) and Na_2O contents of 0.34%.

In the foliated gabbro dike of sample FR16-4B, relic clinopyroxene oc-

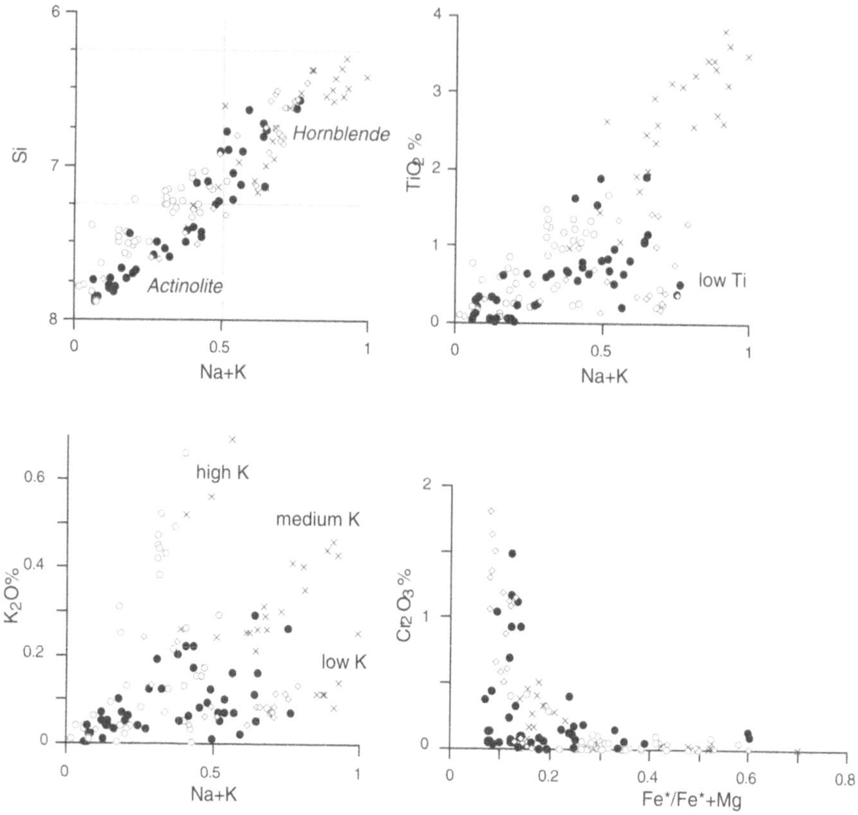

Figure 8. Compositional variations in amphiboles from gabbroic rocks, trondhjemites and veins from the 15°N region. Crosses: brown to green-brown primary amphiboles in gabbroic rocks, trondhjemites and veins; open circles: secondary amphiboles in gabbroic rocks and trondhjemites; closed circles: secondary amphiboles in veins; open diamonds: euhedral colorless amphiboles in veins.

curs in aggregates of anhedral millimeter-sized grains, and as smaller porphyroclasts in the foliated matrix of recrystallized plagioclase and amphibole. Clinopyroxene in aggregates has Mg# between 0.67 and 0.69, Cr_2O_3 contents of 0.1%, Al_2O_3 contents of 1.7 to 1.85%, Na_2O contents of 0.38% and relatively high TiO_2 contents (0.68 to 0.8%). Clinopyroxene porphyroclasts in the gabbro itself have much lower Mg# (0.51 to 0.57; Fig. 7), lower Al_2O_3 and TiO_2 contents (1.3% and 0.45 to 0.62% respectively), and similar Cr_2O_3 and Na_2O contents.

Clinopyroxene in the foliated gabbronorite dike of sample FR22-6 has Mg# ranging from 0.57 and 0.65 in porphyroclasts, and from 0.65 and 0.67 in dynamically recrystallized grains (Fig. 7). Cr_2O_3 contents range from 0.1 and 0.14% in porphyroclasts, and are less than 0.1% in neoblasts. Na_2O, TiO_2 and Al_2O_3 contents are similar in the two types of grains, ranging respectively from 0.36 and 0.45%, 0.3 and 0.5%, and 0.8 and 0.9%.

Feldspar
Anorthite contents in igneous plagioclase from the gabbronorite outcrops of dives FR12 and FR16 range from 53 to 63% (Fig. 7). Plagioclase neoblasts produced during ductile deformation have similar, or only slightly lower anorthite contents, with the exception of An35 plagioclase neoblasts associated with recrystallized actinolite in sample FR12-7 (Fig. 4d).

Gabbro dikes collected in the ultramafics south of the Capo Verde Fracture Zone have plagioclase anorthite contents of 66 to 85% (Fig. 7). The wide range of An values measured in samples FR7-10 and FR9-6 are probably related to the extensive hydrothermal alteration of these samples.

The foliated gabbro dike of sample FR16-4B, contains An42 (Fig. 7) relic igneous plagioclase and dynamically recrystallized plagioclase with lower anorthite contents (11 to 35%) that reflect extensive post-kinematic alteration of these fine-grained neoblasts into actinolite and albitic plagioclase.

Plagioclase in the dike of foliated gabbronorite of sample FR22-6 occurs as neoblasts with anorthite contents of 29 to 32% (Fig. 7). There are no relics of igneous plagioclase porphyroclasts.

Anorthite contents in feldspar from trondhjemitic intrusions range from 0.5 to 46%. The highest values (28 to 46%) are found in mafic xenoliths (Fig. 4e). Plagioclase anorthite contents in the trondhjemitic matrix range from 13 to 30%. In leucocratic veins, albitic plagioclase (An 0.5 to 9.3%) coexists with minor amounts of K-feldspar (0r 78 to 90%).

Amphibole
Brown hornblende of probable magmatic origin occurs in a few gabbroic samples (Table 1) as inclusions in igneous pyroxene or rims around iron-titanium oxides. Relics of similar amphiboles are also found in vein samples. In sample FR21-14, such amphiboles make up most of the vein material. The common characteristic of these brown amphiboles is a high TiO_2 content (2.4 to 3.8%; Fig. 8). Their Na_2O contents range from 1.6 to 3.2% and their maximum chlorine content is 0.16%. Distinctions can be made between these amphiboles, based on their potassium, iron and chromium contents (Fig. 8):

- K_2O contents range from 0.24 to 0.43% in all samples but sample FR21-14, with less than 0.12% K_2O.
- Fe# (Fe_{tot}/Fe_{tot}+Mg) range from 0.14 to 0.53, with the lowest values (0.14 to 0.23) in vein samples, intermediate values (0.27 to 0.30) in gabbronorite samples and the highest values (0.50 to 0.52) in the foliated gabbro dike of sample FR16-4B. An even higher Fe# (0.7) has been measured in magmatic titaniferous hornblende in an olivine-diorite sample from dredge RD88-Dr8 (Cannat et al., 1992).
- Cr_2O_3 contents range from 0 to 0.5%, with low values (0 to 0.05%) in gabbroic samples, and high values (0.17 to 0.5%) in vein samples, except in sample FR21-14 (0.05%).

Green-brown magmatic amphiboles in trondhjemites are actinolitic hornblendes, with 0.9 to 1.5% TiO_2, high Fe# (0.42 to 0.52), 1.2% Na_2O, low Cr_2O_3 (0.05%) and very high K_2O contents (0.3 to 0.7%; Fig. 8). Green-brown euhedral and possibly magmatic amphiboles in vein samples FR16-4C and FR17-4 have higher TiO_2 contents (up to 2.4%), higher Cr_2O_3 contents (up to 0.3%), but a lower K_2O content (0.29%) and much lower Fe# (0.16 to 0.18).

Euhedral colorless amphiboles in veins samples FR5-4, FR9-4, FR19-4 and FR23-5 look similar but have distinct compositions (Fig. 8):
- veins in samples FR5-4 and FR9-4 contain hornblende, with low K_2O contents (0.06 to 0.13%), low Fe# (0.08 to 0.13), low to moderate TiO_2 (0.17 to 1.4%), but very high Cr_2O_3 contents (0.5 to 1.8%).
- veins in samples FR19-4 and FR23-5 are actinolites and actinolitic hornblendes. Sample FR23-5 also contains minor amounts of cummingtonite. These amphiboles have 0.3 to 0.96% TiO_2, low Cr_2O_3 contents (0 to 0.08%), and low to moderate K_2O (0.04 to 0.24%). Fe# are low in sample FR19-4 (0.13 to 0.17), and high in sample FR23-5 (0.43 to 0.60). Calcic and non calcic amphiboles in this sample are similar, although not quite as iron-rich, to secondary amphiboles in dredge RD88-Dr8 diorites (Fe# 0.6 to 0.89; Cannat et al., 1992).

Secondary amphiboles in gabbroic rocks, trondhjemite and veins are brown-green to colorless actinolites or actinolitic hornblendes (Fig. 8). Cummingtonite also occurs in small quantities in the vein of samples FR7-15 and FR16-4A and FR16-4C. Chlorine contents are low (0.01 to 0.39%). In general, the Fe#, K_2O and Cr_2O_3 contents of these secondary amphiboles are similar to that of primary amphibole in the same sample, when present. However, Cr_2O_3 contents in secondary amphiboles from vein samples are commonly higher (up to 1.5% in sample FR10-7, which also contains primary brown amphibole with 0.2% Cr_2O_3). Fe# in secondary amphiboles from vein samples are also commonly lower than in primary amphibole from the same samples.
- secondary amphiboles in gabbronorite samples (Table 1) occur in veins (Fig. 4c) or as in situ replacement of pyroxenes. They are occasionaly recrystallized with fine-grained plagioclase. Their Fe# varies between 0.2 and 0.5. Cr_2O_3 contents are less than 0.1%, TiO_2 contents vary between less than 0.1 and 1.4%. K_2O contents vary between 0.1 and 0.3% in most samples, with higher values (0.5 to 0.7%) in samples FR16-2 and FR16-3 which may be due to contamination of the hydrothermal fluid as it circulated through quartz-biotite veins crosscutting these samples.
- secondary amphiboles in gabbro dikes of samples FR7-10 and FR9-6 have low TiO_2 contents (0.2 to 0.43%) and very low K_2O contents (0.1% or less), even in actinolitic hornblende. Fe# and Cr_2O_3 contents mimic compositional variations of magmatic minerals, with low Fe# (0.08 to 0.2%) and Cr_2O_3 contents as high as 0.41% in sample FR7-10 and higher Fe# (0.39), with no

chromium, in sample FR9-6.
- secondary amphiboles in sample FR16-4B foliated gabbro dike are iron-rich (Fe# 0.40 to 0.58) and chromium-poor (less than 0.1% Cr_2O_3) and comprise dynamically recrystallized greenish-brown actinolitic hornblendes with 0.02 to 0.3% K_2O, and latter acicular actinolite with less than 0.02% K_2O.
- secondary amphiboles in trondhjemite samples are green, potassium-rich actinolites (0.23 to 0.66% K_2O), with high Fe# (0.4 to 0.5), which replace the primary brown-green actinolitic hornblendes.

In vein samples, secondary amphiboles commonly associated with chlorite make up most of the vein filling material. As for the less common brown hornblende, distinctions can be made between these secondary amphiboles based on Fe#, K_2O and Cr_2O_3 contents (Fig. 8).
- high Fe# (0.6) are found in secondary actinolite from sample FR10-9. This value is similar to that of colorless euhedral amphiboles in sample FR23-5, and to that of magmatic and secondary amphiboles in dredge RD88-Dr8 (Cannat et al., 1992). Fe# in secondary amphiboles from other vein samples are low (0.08 to 0.25), to moderate in sample FR7-15 (0.25 to 0.33).
- low K_2O contents (0.01 to 0.12%), high Cr_2O_3 contents (0.5 to 1.5%) and somewhat low TiO_2 contents (0.2 to 1%) are found in secondary actinolitic hornblende from many veins samples, commonly associated with other secondary amphiboles containing more potassium (0.1 to 0.3%) and less chromium. These potassium-poor amphiboles likely resulted from the alteration of pyroxene from xenocrysts and xenoliths of the host ultramafics.

Biotite and Phlogopite

Biotite and phlogopite occur as accessory minerals in some gabbroic and vein samples (Table 1). Biotite is also a major constituent of trondhjemite samples.

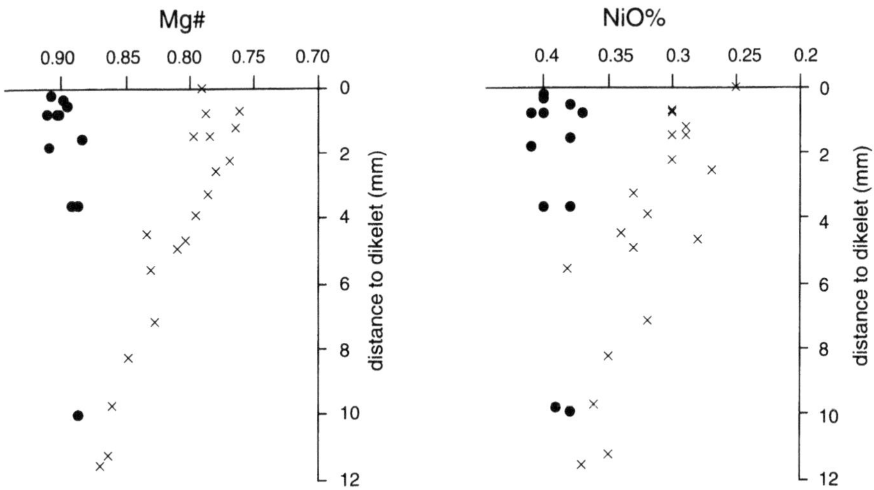

Figure 9. Variations in Mg# and NiO contents of olivine in serpentinized harburgite with distance to veins, in samples FR8-2 (crosses) and FR16-4A and FR23-02 (closed circles).

Biotite in quartz-bearing patches and veins in gabbronorite samples from dives FR12 and FR16, has Fe# between 0.29 and 0.39 and contains a little chlorine (Cl contents of 0.4 to 0.6%). Biotite Fe# are higher in trondhjemites (0.52 to 0.6). Phlogopite occurs in veins samples FR8-2 and FR16-4A. It has low Fe# (0.05 to 0.16) and relatively high NiO contents (0.2 to 0.3%).

Chemical modifications of host ultramafics near veins and dikes

Systematic investigation of chemical interactions between dikes, veins and their host ultramafics is possible in a limited number of samples only, due to incomplete sampling and to extensive serpentinization. However, many dike or vein-bearing ultramafic samples contain at least one relic of spinel, olivine or orthopyroxene within a few centimeters of the dike or vein margins. When located at distances greater than about 2 cm from these margins, these minerals are always compositionally similar to those of ultramafics collected away from intrusions (Fig. 6). The few relatively less serpentinized peridotite samples in which relic primary minerals are preserved within 2 cm from dikes or veins margins belong to two groups (Fig. 9):
(1) Olivine, spinel and orthopyroxene in serpentinized harburgites within 2 cm from the foliated gabbronorite dike of sample FR22-6, and from the veins of samples FR8-2 and FR8-13 are chemically modified. Olivine Mg# and NiO contents decrease towards the dike margins, with minimum Mg# of 0.7 to 0.75 (against 0.91 to 0.92 in peridotites collected away from intrusions; Fig. 6), and minimum NiO contents of about 0.25% (against 0.34 to 0.41% in Fig. 6). Orthopyroxene Mg# follows the same trend (minimum values of about 0.8), with no change in aluminium contents. Spinel Mg# can be as low as 0.15 at dike margins, while Cr# increases to about 0.7. Spinels may also be enriched in titanium, with TiO_2 contents as high as 2.7% near the vein in sample FR8-13. But such titanium enrichments are not systematic (no enrichment in sample FR22-6 dike's margin). Olivine and clinopyroxene in sample FR7-15 serpentinized wehrlite are also enriched in iron near the zircon-bearing vein (Fig. 4a), with Mg# as low as 0.78 in olivine, and 0.88 in clinopyroxene. There is no associated decrease of olivine NiO contents, possibly because they are already similar (about 0.25% in olivine away from the vein; Fig. 6) to NiO contents in the most chemically modified olivine grains of Fig. 9.
(2) Olivine, spinel and orthopyroxene in serpentinized harburgites within 2 cm from the veins of samples FR5-4, FR17-4, FR16-4C and FR23-2 are not chemically modified.

Discussion

Ca-rich gabbroic rocks: products of a high degree of mantle melting?

Mineral compositions in massive gabbronorites from dives FR12 and FR16 north of the Capo Verde Fracture Zone are very similar to those of olivine and oxide-bearing gabbros drilled at ODP Site 735B (Ozawa et al., 1991; Bloomer et al., 1991), and to other evolved gabbros dredged in the Southwest Indian Ocean (Bloomer et al., 1989). In the clinopyroxene Mg# versus plagioclase An% diagram of Fig. 7, clinopyroxene and recrystallized plagioclase (no relic igneous plagioclase in this sample) in the foliated gabbronorite dike of sample FR22-6 plot near the most evolved oxide gabbros from ODP site 735B. Based on compositions of igneous and of similar medium-sized recrystallized plagioclase measured in other deformed gabbronorites, it is safe to assume that dynamic recrystallization did not significantly alter plagioclase An contents in this sample. Recrystallized clinopyroxene in this dike is enriched in magnesium (error bar in Fig. 7), compared to relic igneous porphyroclasts. Sub-solidus iron-magnesium exchanges with the surrounding harzburgite during ductile deformation can account for this magnesium enrichment.

In contrast, the other gabbroic dikes collected in the 15°N region are offset to higher plagioclase An contents, compared with the Southwest Indian Ridge sample set (Fig. 7). The three gabbro dikes collected in serpentinized ultramafics south of the Capo Verde fracture zone actually plot close to Ca-rich gabbros from DSDP site 334 (Mid-Atlantic Ridge near the Azores; Tiezzi and Scott, 1980; Ross and Elthon, 1993) or from the Mariana arc (Bloomer, personal communication 1993). Two of these three southern gabbro dikes also have primitive compositions, with high clinopyroxene Cr203 contents and Mg# plotting close to those of sample FR7-15 wehrlite. The occurrence of accessory zircon in sample FR7-10 (Table 1) does not quite fit with such primitive compositions. Zircon occurs, however, in the most extensively altered and deformed part of the sample, and our preferred explanation is that it does not belong to the gabbro dike, but to a later vein crosscutting it and now obscured by shearing and hydrothermal alteration. Apatite and ilmenite are unambiguously accessory minerals in sample FR9-6, consistent with the relatively low Mg# and Cr_2O_3 contents of clinopyroxenes in this sample.

Gabbros collected as dikes in the ultramafics south of the Capo Verde transform may thus have derived from primary melts with lower sodium contents than parent melts of the northern gabbronorites. This may reflect higher degrees of partial melting in the mantle beneath the southern region. Olivine, orthoyroxene and spinel compositions (Fig. 6) of serpentinized harzburgites from this southern region are consistent with this interpretation, being the most depleted ever sampled along the Mid-Atlantic Ridge. Basalt studies, showing N-type MORBs north of the Capo Verde Fracture Zone, and E-type MORBs to the south (Bougault et al., 1988; Xia and Casey, 1991; Dosso et al., 1993), also support this hypothesis.

The foliated gabbro dike in sample FR16-4B, and the olivine-diorite collected in dredge RD88-Dr8 (Cannat et al., 1992), also plot on a Ca-rich trend in Fig. 7. Plotting clinopyroxene from aggregates (Mg# 0.67 to 0.69) against plagioclase porphyroclasts in sample FR16-4B would put us back into the SWIR trend, but would make no sense as clinopyroxenite aggregates in this sample are relics of an earlier mineral assemblage. The petrogenesis of these iron-rich intrusives needs to be discussed in the light of trace and REE data (Casey et al., in prep.). They may be evolved end members of sodium-poor primary melts, suggesting that melts derived from high degrees of mantle melting were also involved in building the crust north of the Capo Verde transform. Alternatively, they could result from fractionation processes of the kind discussed in the next paragraph, starting with melts issued from moderate to low degrees of mantle melting.

Trondhjemites and quartz-bearing veins: products of mechanical melt segregation?

Trondhjemites are found as meter-sized intrusions in serpentinized peridotites north of the Capo Verde transform. There is actually only one other published account of such acidic rocks at a mid-ocean ridge (Engel and Fisher, 1975). The 15°N trondhjemites are fine-grained and were therefore probably emplaced at upper lithospheric levels, in ultramafics that had already cooled substantially. Quartz-biotite veins are also found in some gabbronorite samples (Fig. 4b and c), where they crystallized during the last stages of brittle-ductile deformation. We propose that these synkinematic quartz veins are filled with interstitial melts extracted out of the gabbronorites crystalline matrix. These interstitial melts, representing small volumes of magma left over in isolated pockets at the end of fractional crystallization, could have a variety of evolved compositions. Mechanical segregation during deformation of the host gabbroic material could cause such melts to pool to form intrusions of trondhjemite, and possibly also of other highly evolved lithologies such as diorites. A similar process was proposed for the formation of highly differenciated oxide-rich gabbros in ODP site 735B drilled section (Dick et al., 1991; Bloomer et al., 1991).

Nature of veins protoliths

Most veins sampled in the ultramafics both north and south of the Capo Verde transform have a probably magmatic origin. These veins contain zircon, ilmenite, titanium-rich primary amphibole and pseudomorphs after presumably igneous pyroxene (Table 1). Other veins, made mostly of euhedral colorless amphibole (samples FR5-4, FR9-4, FR19-4 and FR23-5), are probably hydrothermal. The occurrence of accessory ilmenite in sample FR23-5 suggests, however, that prismatic amphibole there replaced an earlier igneous assemblage. High chromium contents in colorless amphibole from samples FR5-4 and FR9-4 are consistent with these veins representing hydrothermal replacements after primitive pyroxenitic dikelets in the ultramafics. Finally, veins from samples FR10-4, FR19-1 and FR23-4, lacking these distinctive features, are of uncertain origin.
Accessory minerals such as zircon, ilmenite, and apatite suggest that altered,

presumably magmatic, veins in the 15°N ultramafics had evolved igneous compositions. As other relic primary silicate phases are lacking, further discussion of compositional characteristics of these veins is mostly based on amphibole chemistry. This is relatively straightforward in samples containing relic primary amphibole as an accessory mineral (samples FR8-13, FR10-7 and FR21-6), or as a major phase (samples FR17-4 and FR21-14). In all samples, evidence for chemical exchanges between veins and their host ultramafics must, however, be taken into account. These exchanges are probably responsible for the relatively low iron and hich chromium contents of most primary vein amphiboles.

Veins from samples FR8-13, FR10-7 and FR21-6 are texturally similar (judging from altered pseudomorphs after primary minerals) to the foliated gabbronorite dike of sample FR22-6. Accessory mineralogy and the chemistry of primary and secondary amphiboles are consistent with this interpretation. Lower iron contents in the veins amphiboles are attributed to chemical interactions with host ultramafics. The abundant titanium-rich hornblende in sample FR21-14 is also iron-poor, with similar titanium contents, but significantly lower potassium contents (Fig. 8). The origin of such low potassium contents in primary amphiboles is not understood, but probably reflects differences in compositions for the melt that crystallized in this sample.

Veins in samples FR16-4C and FR17-4 contain abundant ultramafic xenoliths (Fig. 5) and may represent magmatic breccias at the contact between trondhjemite intrusions and the surrounding ultramafics. The low iron, high chromium and nickel contents measured in primary greenish-brown actinolitic hornblende from sample FR17-4 is inferred to result from chemical interactions between the melt and the ultramafics. Primary minerals in xenoliths of serpentinized harzburgite in this sample are, however, not chemically modified, suggesting that melt rock interactions occurred at temperatures too low for these minerals to reequilibrate. Such low temperatures are consistent with the fine-grained texture of this vein and with the crystallization of actinolitic primary amphiboles. We propose that veining coincided with hydrous alteration of the peridotite into a talc-bearing assemblage. Melt contamination may then have occurred through partial assimilation of this altered assemblage, producing the intricate talc-amphibole mesh, with accessory phlogopite, zircon or sulphides, also observed in sample FR16-04A.

In samples lacking primary amphiboles, inferences may be drawn from the composition of secondary amphiboles. In a given sample, compositional variations between primary and secondary amphiboles are consistent with chemical interactions between hydrothermal fluids and the surrounding ultramafics: secondary amphiboles tend to have lower titanium and potassium contents, higher chromium and nickel contents, but similarly low, or even lower iron contents than primary amphiboles. High Fe# (0.6) measured in secondary actinolite from sample FR10-9 may therefore reflect high Fe# in the original igneous minerals of this vein. The same may apply to similarly iron-rich amphiboles in sample FR23-5. Similar iron contents have been measured in primary and secondary amphiboles from the foliated gabbro

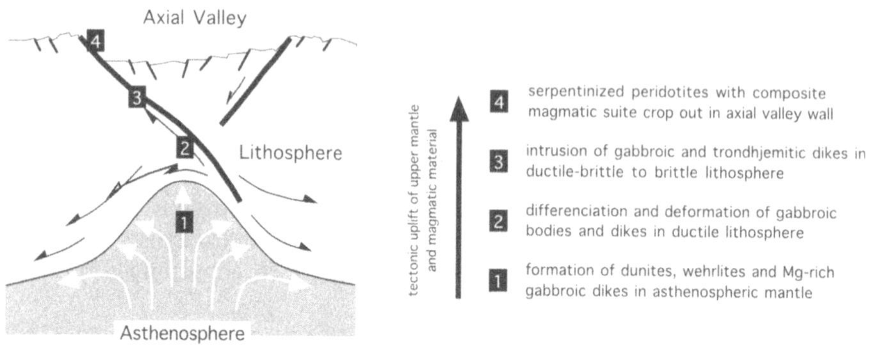

Figure 10. Sketch of a magma-starved oceanic ridge, with successive stages of magmatism within mantle-rocks rising up from the asthenosphere and through the axial lithosphere to eventually form seafloor exposures. See text for detail.

dike of sample FR16-4B and from olivine-diorite intrusives collected in dredge RD88-Dr8 (Cannat et al., 1992). Samples FR10-9 and FR23-5, coming from very near this dredge, may represent altered veins of this dioritic melt.

Deformation history

The sequence of deformational events observed in the gabbroic rocks from the 15°N region is similar to that of ODP Site 735B gabbroic section (Cannat, 1991; Cannat et al., 1991) and include (1) oriented magmatic flow producing shape and crystallographic fabrics with no intracrystalline deformation, (2) ductile deformation and recrystallization of magmatic minerals, including orthopyroxene where present (granulite to upper amphibolite facies metamorphic conditions), into relatively large (0.2-0.3 mm) polygonal neoblasts, and (3) brittle-ductile deformation with recrystallization of plagioclase and of hydrothermal actinolitic hornblende to actinolite (lower amphibolite to greenschist facies metamorphic conditions) into small (0.04 mm or less) irregularly-shaped neoblasts, and brittle failure of plagioclase and pyroxene porphyroclasts (Fig. 4c). The decrease of neoblast sizes between stages 2 and 3 may be interpreted as an indication of increased flow stresses, consistent with the expected yield strength increase as temperature decreased from granulite to greenschist facies metamorphic conditions.

Ultramafics north of the Capo Verde transform are strongly foliated, with ribbon-shaped orthopyroxene and extensively recrystallized olivine. Microstructures and fabrics created during this deformation suggest that it occurred in conditions thought to prevail in the lower axial lithosphere (Cannat et al., 1992). This inferred lower lithospheric event also affected the foliated gabbronorite dike of sample FR22-6, causing partial recrystallization of igneous minerals (pyroxenes, plagioclase and accessory apatite, amphibole and iron-titanium oxides) into neoblasts 0.1 to 0.3 mm in size (stage

2 deformation event as defined above).

Ultramafics collected south of the Capo Verde transform have weakly defined foliations, large grain sizes, and cuspate to straight grain boundaries with common 120° triple junctions. These textures suggest deformation at solidus or close to solidus temperatures and relatively low deviatoric stresses, i.e., asthenospheric conditions. These ultramafics clearly had to rise through the axial lithosphere in order to reach their present position in the seafloor. Their lack of a lithospheric deformation imprint is an indication that extensional deformation in the axial lithosphere was localized into discrete shear zones (Fig. 10). The ultramafics outcrops north of the Capo Verde transform may provide samples of such lower lithospheric shear zones.

Successive stages of magmatism

Magmatic, textural and deformational constraints outlined above suggest that the 15°N ultramafics recorded successive stages of magma emplacement, under progressively lower temperature conditions which can be schematically ascribed to progressively lower depths beneath the axial seafloor (Fig. 10). As all ultramafic, gabbroic and trondhjemitic samples described in this paper were recovered from the seafloor, these successive stages of magma emplacement must be envisioned as accompanying the uplift of mantle material from the sub-axial asthenosphere, through the lithosphere. There is no geophysical data in the 15°N region to help us constrain axial lithospheric thickness. A parallel may, however, be drawn on the basis of similar axial valley relief and ultramafic exposures, with the 23°N MARK (Mid-Atlantic Ridge/Kane fracture zone) area, where microearthquakes with focal depths down to 8 km have been recorded in the axial domain (Toomey et al., 1988). These maximum focal depths are thought to correspond with the brittle-ductile transition in the axial lithosphere (Toomey et al., 1988). If this interpretation is valid, the base of the ductile lithosphere would be deeper (of the order of 10 km or more). The picture of the axial distribution of magmatic rocks as it is drawn in Fig. 10 from observations discussed in this paper, is very different from the more traditional, layered view of the oceanic crust (Penrose, 1972) recently refined on the basis of results of seismic experiments along the world ridge system (Sinton and Detrick, 1992). Possibly the most significant difference between the two models is that the layered crust model views the axial lithosphere-asthenosphere boundary as coincident with the roof of the axial magma chamber or magma lense whilst, in the magma-starved environment depicted in Fig. 10, gabbroic intrusions occur in a discontinuous fashion throughout the lithosphere, within mantle rocks that are continuously tectonically unroofed to eventually form seafloor exposures.

Stage 1 magmatic events include the formation of dunites and wehrlites in the asthenospheric mantle sampled south of the Capo Verde transform, the probable impregnation of residual harzburgites by small volumes of interstitial melts, and the crystallization of coarse-grained dikes of primitive gabbros. Similar olivine and spinel composition in serpentinized dunites and harzburgites collected north of the Capo Verde transform suggest a common residual origin. By contrast, spinel and olivine composition (Fig. 6) in serpentinized dunites sampled south of the Capo Verde transform point to a magmatic origin, or to pronounced interactions of residual ultramafics with basaltic melts (Quick, 1981). Mineralogically and texturally, these southern dunites and the wehrlites are similar to those of mantle/crust transition zones exposed in many an ophiolite (Casey et al., 1981; Nicolas and Prinzhofer, 1983; Benn et al., 1988), and probably formed through similar magmatic and melt/mantle interaction processes.

Stage 2 magmatic events include the emplacement and differenciation of gabbroic rocks in ultramafics and in stage 1 magmatic rocks after they were incorporated into the lower axial lithosphere. Coarse to medium grained gabbronorites collected north of the Capo Verde transform were emplaced at this stage, as well as dikes and veins of evolved gabbros and diorites. The extent of magmatic differentiation recorded in these second stage intrusives suggest that they did not crystallize in frequently replenished, long-lived magma bodies, but rather in short-lived dikes or sills. Ductile deformation in granulite to upper amphibolite facies conditions, and limited chemical exchanges between gabbroic rocks and their host ultramafics occurred at this stage. A rough evaluation of the volume of ultramafics that were chemically modified at this stage may be attempted, based on samples collected north of the Capo Verde transform. Veins-bearing samples represent 40% of the ultramafics collected in this northern area. We assume that this proportion is representative of actual outcrops, which is far from proven, and that 50% of these samples contain veins emplaced in the lower lithosphere, that produced chemical exchanges over distances no more than 2 cm from the veins margins. This leads to no more than 2% in volume of the mantle rocks exposed in the northern area being chemically modified during stage 2.

Stage 3 magmatic events developed higher up in the axial lithosphere, in ductile-brittle, to brittle conditions. These events included mechanical segregation of evolved late magmatic melts in gabbroic rocks emplaced at stage 2, the intrusion of trondhjemites, and the formation of veins and dikes in ultramafics that had cooled to temperatures too low for chemical exchanges to be effective. Rare diabase and basalt dikes (Casey et al., in prep.), cutting the ultramafics and presumably acting as feeder dikes for axial valley floor basalt flows, were also emplaced at this stage. Veining in some cases may have coincided with hydrous replacement of the host peridotite by a talc, actinolite and chlorite-bearing greenschist facies assemblage. However, in all samples, igneous veins and dikes are cut by serpentine veins, suggesting that they predate extensive serpentinization of the ultramafics.

Acknowledgements:

The rocks studied in this paper were all collected during the FARANAUT diving cruise (R/V l'Atalante and Nautile) in march 1992. Our thanks go to the captain and crew and to our fellow scientists (H. Bougault who was chief scientist, P. Appriou, J.L. Charlou, L. Dmitriev, Y. Fouquet, P. Jean Baptiste and P. Rona). Our work received financial support from CNRS-INSU (IST-Géosciences Marines).

References

Benn, K., Nicolas, A. and Reuber, I., 1988. Mantle-crust transition zone and origin of wehrlitic magmas: evidence from the Oman ophiolite. Tectonophysics, 151: 75-85.

Bloomer, S.H., Natland, J.H. and Fisher, R.L., 1989. Mineral relationships in gabbroïc rocks from fracture zones of Indian Ocean ridges: evidence for extensive fractionation, parental diversity, and boundary-layer recrystallization. In: A.D. Saunders and M.J. Norry (Editors), Magmatism in the Ocean Basins. Geol. Soc. London Spec. Publ., 42: 107-124.

Bloomer, S.H., Meyer, P.S., Dick, H.J.B., Ozawa, K. and Natland, J.H., 1991. Textural and mineralogic variations in gabbroic rocks from hole 735B. In: R.P. von Herzen, P.T. Robinson, et al., Proc. ODP, Sci. Results, 118: College Station, TX (Ocean Drilling Program), 21-39.

Bonatti, E., Seyler, M. and Sushevskaya, N., 1993. A cold suboceanic mantle belt at the Earth's Equator. Science, 261: 315-320.

Boudier, F., Le Sueur, E. and Nicolas, A., 1989. Structure of an atypical ophiolite: the Trinity Complex, eastern Klamath Mountains, California. Geol. Soc. America Bull., 101: 820-833.

Bougault, H., Charlou, J.L., Fouquet, Y., Needham, H.D., Vaslet, N., Appriou, P., Jean Baptiste, P., Rona, P.A., Dmitriev, L. and Silantiev, S., 1993. Fast and slow spreading ridges: structure and hydrothermal activity, ultramafic topographic highs and CH4 output. J. Geophys. Res., 98: 9643-9651.

Cannat, M., Mével, C. and Stakes, D.S., 1991. Stretching of the deep crust at the slow spreading Southwest Indian Ridge. Tectonophysics, 190: 73-94.

Cannat, M., 1991. Plastic deformation at an oceanic spreading ridge: a microstructural study of the site 735 gabbros, Southwest Indian Ocean, Leg ODP 118. In: R.P. von Herzen, P.T. Robinson, et al., Proc. ODP, Sci. Results, 118: College Station, TX (Ocean Drilling Program), 399-408.

Cannat, M., Bideau, D., and Bougault, H., 1992. Serpentinized peridotites and gabbros in the Mid Atlantic Ridge axial valley at 15°37'N and 16°52'N. Earth Planet. Sci. Lett., 109: 87-106.

Cannat, M., 1993. Emplacement of mantle rocks in the seafloor at mid-ocean ridges. J. Geophys. Res., 98: 4163-4172.

Casey, J.F., Dewey, J.F., Fox, P.J., Karson, J.A. and Rosencrantz, E., 1981. Heterogeneous nature of oceanic crust and upper mantle. A perspective from the Bay of Islands ophiolite complex. In: C. Emiliani (Editor), The Sea., Wiley, New York, 305-338.

Dick, H.J.B., 1989. Abyssal peridotites, very slow spreading ridges and ocean ridge magmatism, In: A. D. Saunders and M. J. Norris (Editors), Magmatism in the Ocean Basins. Geol. Soc. London Spec. Publ., 42: 71-105.

Dick, H.J.B., Meyer, P.S., Bloomer, S.H., Kirby, S., Stakes, D., and Mawer, C., 1991. Lithostratigraphic evolution of an in-situ section of oceanic layer 3. In: R.P. von Herzen, P.T. Robinson, et al., Proc. ODP, Sci. Results, 118: College Station, TX (Ocean Drilling Program), 439-538

Dosso, L., Bougault, H. and Joron, J.L., 1993. Geochemical morphology of the North Atlantic Mid-ocean Ridge: 10°-24°N. Trace element-isotope complementarity. Earth Plan. Sci. Lett., 120: 443-462.

Engel, C.G. and Fisher, R.L., 1975. Granitic to ultramafic rock complexes of the Indian Ocean ridge system, Western Indian Ocean. Geol. Soc. America Bull., 86: 1553-1578.

Girardeau, J. and Francheteau, J., 1993. Plagioclase-wehrlites and peridotites on the East Pacific Rise (Hess Deep) and Mid-Atlantic Ridge (DSDP Site 334): evidence for magma percolation in the oceanic upper mantle. Earth Plan. Sci. Lett., 115: 137-149.

Karson, J.A., 1991. Seafloor spreading on the Mid-Atlantic Ridge: Implications for the structure of ophiolites and oceanic lithosphere produced in slow-spreading environments. In: J. Malpas, E. M. Moores, A. Panayiotou and C. Xenophontos (Editors), Proceedings of the Symposium "Troodos

1987", Geol. Survey Department, Nicosia, Cyprus, 547-555.
Karson, J. A., Thompson, G., Humphries, S.E., Edmond, J.M., Bryan, W.B., Brown, J.R., Winters, A.T., Pockalny, R.A., Casey, J.F., Campbell, A.C., Klinkhammer, G., Palmer, M.R., Kinzler, R.J. and Sulanowska, M.M., 1987. Along axis variations in seafloor spreading in the MARK area. Nature, 328: 681-685.
Kushiro, I. and Yoder, H.S., 1966. Anorthite-forsterite and anorthite-enstatite reactions and their bearing on the basalt-eclogite transformation. J. Petrol., 7: 337-362.
Lagabrielle, Y. and Cannat, M., 1990. Alpine Jurassic ophiolites resemble the modern Central Atlantic basement. Geology, 18: 319-322.
Mercier, J.C., and Nicolas, A., 1975. Textures and fabrics of upper mantle peridotites as illustrated by xenoliths from basalts. J. Petrol., 16: 454-487.
Mével, C., Cannat, M., Gente, P., Marion, E., Auzende, J.M. and Karson, J.A., 1991. Emplacement of deep rocks on the west median valley wall of the MARK area (Mid-Atlantic Ridge 23°N). Tectonophysics, 190: 31-53.
Meyer, P.S., Dick, H.J.B. and Thompson, G., 1989. Cumulate gabbros from the Southwest Indian Ridge, 54°S-7°16'E: Implications for magmatic processes at a slow spreading ridge. Contrib. Mineral. Petrol., 103: 44-63.
Nicolas, A. and Prinzhofer, A., 1983. Cumulative or residual origin for the transition zone in ophiolites: structural evidence. J. Petrol., 24: 188-206.
Ozawa, K., Meyer, P. and Bloomer, S., 1991. Mineralogy and textures of iron-titanium oxide gabbros and associated olivine gabbros from hole 735B. In: R.P. von Herzen, P.T. Robinson, et al., Proc. ODP, Sci. Results, 118: College Station, TX (Ocean Drilling Program), 41-73.
Penrose, 1972. Ophiolites, Penrose Field Conference. Geotimes, 17: 24-25.
Quick, J.E., 1981. The origin and significance of large, tabular dunite bodies in the Trinity peridotite, northern California. Contrib. Miner. Petrol., 78: 413-422.
Ross, K. and Elthon, D., 1993. Cumulates from strongly depleted mid-ocean ridge basalt. Nature, 365: 826-829.
Sinton, J.M. and Detrick, R.S., 1992. Mid-ocean ridge magma chambers. J. Geophys. Res., 97: 197-216.
Tiezzi, L.J. and Scott, R.B., 1980. Crystal fractionation in a cumulate gabbro, Mid-Atlantic Ridge, 26°N. J. Geophys. Res., 85: 5438-5454.
Toomey, D.R., Solomon, S.C., Purdy, G.M. and Murray, M.H., 1988. Microearthquakes beneath the median valley of the Mid-Atlantic Ridge near 23°N: Tomography and tectonics. J. Geophys. Res., 93: 9093-9112.
Xia, C., Casey, J.F., Silantiev, S. and L. Dmitriev, 1991. Geochemical structure of the 14 N mantle source anomaly along the Mid-Atlantic Ridge and geochemical changes across the 15° 20' N Fracture Zone. (Abstract) EOS, Trans. Am. Geophys. Union, 72: 518.

Gabbroic Dikelets in Serpentinized Peridotites from the Mid-Atlantic Ridge at 23°20'N

P. TARTAROTTI, M. CANNAT* AND C. MEVEL*

Dipart. di Geologia, Paleontologia e Geofisica, Università di Padova, Via Giotto 1, 35137 Padova, Italy

** Laboratoire de Pétrologie, Université de Paris VI, 4 Place Jussieu, 75251 Paris Cedex 05, France*

Abstract

Mantle-derived peridotites sampled on the seafloor at 23° N along the Mid-Atlantic Ridge during the Hydrosnake and R.V. Akademic Mstislav Keldysh cruises show evidence of minor magmatic intrusion that produced millimetre to centimetre thick gabbroic dikelets. The nature of the intruding magma and the time relationships between intrusion events and tectonic evolution of the host peridotites is inferred from the mineralogy and textural features of veins cutting the peridotites. The occurrence of Fe-Ti-rich minerals and of zircon in the veins suggests an evolved, basaltic composition of the fluid. Pyroxene-rich, plagioclase-free veins probably represent pyroxenite dikelets inside the peridotites. In all cases, melt intrusion followed ductile deformation affecting the mantle-derived ultramafics. These observations fit with a model in which gabbroic intrusions are emplaced in mantle rocks at shallow depths beneath the ridge axis.

Introduction

The occurrence, along the Mid-Atlantic Ridge, of plutonic rocks of the lower crust and upper mantle in tectonic settings away from fracture zone walls is well documented (Phillips et al., 1968; Aumento and Loubat, 1971; Bonatti et al., 1975; Melson, Rabinowitz et al., 1975; Tiezzi and Scott, 1980; Bougault, Cande et al., 1985; Karson et al., 1987; Rona et al., 1987; Detrick, Honnorez et al., 1988; Mével et al., 1988, 1991; Zonenshain et al., 1989; Bougault et al., 1990). In the MARK area (Mid-Atlantic Ridge at the Kane Transform), outcrops of plutonic rocks have been observed and sampled between 23°N and 24°N during DSDP Leg 45 and ODP Leg 109 (Melson, Rabinowitz et al., 1975; Detrick, Honnorez et al., 1988), during Alvin (Karson and Dick, 1983; Karson et al., 1987) and Nautile dives (Mével et al., 1988, 1991), and by the Soviet submersibles (Gente et al., 1989).

These occurrences suggest an anomalous oceanic crust which contrasts with the classical model of a layered crust consisting of volcanic extrusives, sheeted dikes and gabbros. Such a situation is the basis of many speculative

R.L.M. Vissers and A. Nicolas (Eds.),
Mantle and Lower Crust Exposed in Oceanic Ridges and in Ophiolites, 35-69.

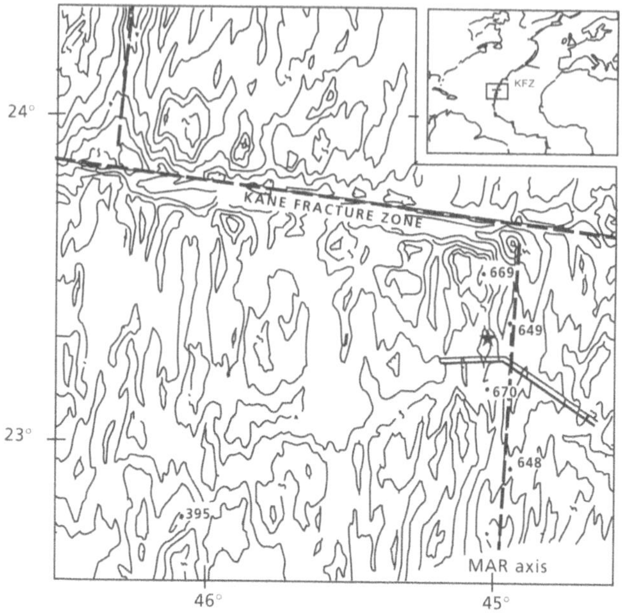

Figure 1. Bathymetric map of the MAR/Kane fracture zone intersection (MARK area), after Tucholke and Schouten (1988). Solid circles are DSDP and ODP drill sites. Star is the peridotite-bearing hill culminating at 2600 mbsl (see text). Parallel bold lines mark the discordant zone (non transform offset; Schulz et al., 1988) between the northern and southern spreading cells (after Purdy and Detrick, 1986).

tectonic interpretations (e.g. Karson, 1991; Mével et al., 1991; Francis, 1981; Cannat, 1993; see also discussion below). In order to shed light on the tectonic evolution of the MARK area we studied some mantle derived peridotites, sampled at 23° N in the western axial valley wall during the Hydrosnake cruise (Mével et al., 1988, 1991) and during the R.V. Akademic Mstislav Keldysh cruise (Gente et al., 1989). These peridotites are cut by isolated, centimetre thick altered veins. The mineral assemblages, compositions and textural features allow us to infer the nature of these veins. Microstructural relationships are also used to constrain the timing of the intrusion of the veins relative to the structural and metamorphic evolution of the surrounding mantle rocks. We finally discuss how these events fit into a crust-mantle evolutionary model.

Geological setting

The MARK area (Fig. 1) is one of the most extensively surveyed portions of the Mid-Atlantic Ridge. A detailed bathymetric SeaBeam map of the ridge segment lying between the Kane transform (23°35'N) and 22°30'N was produced during the survey preliminary to Leg 106 (Detrick et al., 1984; Kong et al., 1988; Pockalny et al., 1988). It shows the existence of a relatively linear, about North-South trending rift valley which varies in width from 10 to 17 km. The inner valley floor is shallowest, i.e., less than 3200 m below sea level (bsl) about 70 km south of the Kane Transform. It gradually deepens northward, plunging to depths of about 6000 m in the ridge-transform intersection area (nodal basin). The inside corner of the nodal basin is characterized by a strong topographic high (1300 mbsl) while the outside corner is much lower (3500 mbsl). This asymmetry is a usual feature of slow-spreading ridge-fracture intersection zones (Fox and Gallo, 1984; Severinghaus and Macdonald, 1988).

In the MARK area, two domains with contrasting seismic structure have been distinguished (Fig.1): a northern cell extending from the transform-ridge intersection down to 23°18'N, with a seismic crustal thickness of 4-5 km, and a southern cell with a seismic crust 6-7 km thick. The contrast in crustal thickness between the two cells is also clearly shown by gravity data (Morris and Detrick, 1991). The transition between these two cells corresponds with a zone of low magnetic intensities, and to a probable minor offset of the magnetic anomalies (Schulz et al., 1988). Crustal thinning towards the transform-ridge intersection in this area is also supported by previous seismic refraction experiments (Cormier et al., 1984) which infer a crustal thickness of only 1-2 km below the nodal basin. Plutonic rock exposures in the MARK area, locally overlain by altered basalts, are concentrated near the nodal basin and close to the transition between the northern and southern spreading cells. This occurrence of deep crustal and mantle rock outcrops has been interpreted as a result of extreme stretching and thinning of the crust (Karson and Dick, 1983; Karson et al., 1987). During the Alvin cruise, two serpentinite outcrops were discovered on the western wall of the axial valley (Karson et al., 1987): one at 23°10'N, the other at 23°21'N. The southern outcrop was drilled during Leg 109 (Site 670, Detrick, Honnorez, et al., 1988; see Fig. 1). The results presented in this paper concern the northern outcrop which was dredged during a cruise of the R.V. Akademik Mstislav Keldysh (Gente et al., 1989) and explored during two Nautile dives (Mével et al., 1991). This northern outcrop forms a hill culminating at 2600 mbsl (Figs. 1, 2). Pillow basalts crop out at the bottom of the hill, in the median valley floor. The base of the slope, at about 3700 mbsl, is a talus made of serpentinite and basalt fragments, in some places semi-consolidated and subsequently furrowed by small gullies which indicate recent reactivation of the slope (Mével et al., 1991). The serpentinized peridotites crop out quite continuously from about 3500 mbsl, to about 3100 mbsl. These outcrops commonly display a schistose appearance, probably due to numerous irregular serpentinite veins and fractures, which dip 20° to 50° to the east. There are also fault surfaces, up to several centimetres thick, dipping 40° to 70° to the east, with down-dip striations and presumably normal senses of offset. The peridotites are also cut by numerous, 1 to 10 metre scale fractures with highly variable orientations and frequent serpentinite slickensides. Above 3100 m, talus made of serpentinite and basalt fragments conceals the contact between the ultramafic rocks and the pillow basalts which form the top of the hill. The pillows may be stratigraphically overlying the peridotites. Alternatively, basalts and peridotites may belong to two tectonically juxtaposed blocks. The pillows are cut by numerous steep fault scarps, the highest of which faces east and is about 250 m high.

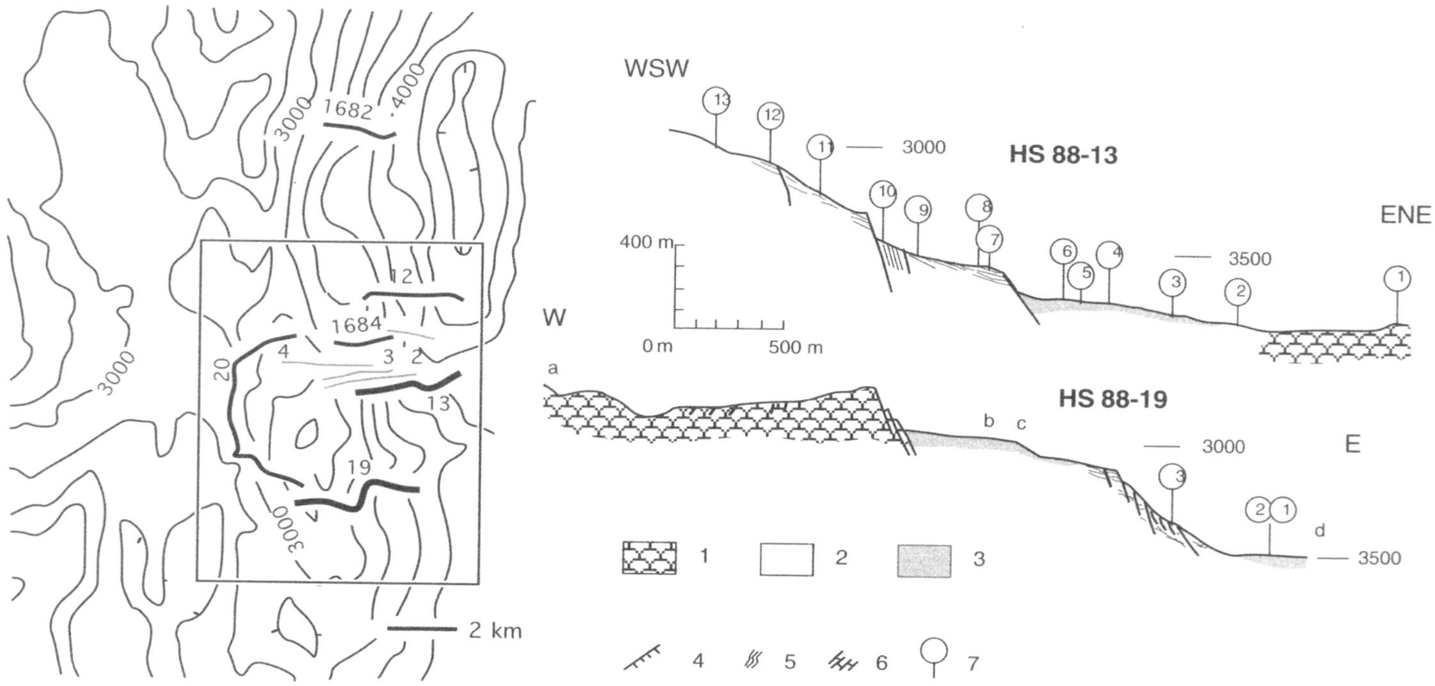

Figure 2. Interpretative cross sections and sample locations along the western valley wall near the peridotite-bearing hill, as established from Nautile dives (after Mével et al., 1991). Bold lines: Nautile dives; grey lines: dredgings of Akademik Mstislav Keldysh. Contours at 100 m intervals. (1) pillow-basalt; (2) serpentinized peridotite; (3) talus; (4) fault; (5) schistosity; (6) joints; (7) sample number. Note that subsurface extent of rock units is unknown, and that ornaments do not necessary reflect thicknesses of rock units.

Petrographic description

The serpentinized ultramafics studied include samples which are cross-cut by isolated millimetre to centimetre scale veins. Sample locations on the eastern flank of the peridotite-bearing hill are shown in the cross-sections of Fig. 2. Brief petrographic descriptions of the samples are listed in Table 1.

Textural characteristics and deformation of the serpentinized ultramafics

The ultramafic rocks are extensively serpentinized, coarse-grained tectonites which exhibit a porphyroclastic texture (Mercier and Nicolas, 1975). The foliation is marked by the shape fabric of orthopyroxene porphyroclasts and spinel, but it is not always well defined owing to the equant dimensions of orthopyroxene. Modal compositions of the serpentinized peridotites may be reconstructed with the following assumptions (Juteau et al., 1990): the mesh-texture and magnetite in the serpentine matrix are counted as olivine, bastite pseudomorphs are assigned to primary orthopyroxene, magnetite rimming the Cr-spinel is counted as primary spinel, and no volume increase is considered. It follows that the modal calculations reported here may have significant uncertainties in the relative proportion of the primary phases. Based on the above assumptions, the average modal composition of the studied rocks is 81% olivine, 14% orthopyroxene, 3.2% clinopyroxene and 1.8% spinel. These rocks may therefore be classified as serpentinized harzburgites. One sample (DR1-2A) dredged with the Akademik Mstislav Keldish consists of serpentinized olivine and spinel, and is therefore a serpentinized dunite. Although they are highly altered, we refer in this study to the rocks in terms of their inferred primary mineralogy.

Olivine has mostly been observed as isolated crystals in the serpentine network. In a few cases, millimetre-sized olivine porphyroclasts with subgrain boundaries have been preserved. They are rimmed by dynamically recrystallized grains, about 20 μm in size.

Orthopyroxene forms millimetre to centimetre-scale, elongate or ovoid porphyroclasts, usually replaced by yellow-brown fibrous bastite pseudomorphs which locally preserve pyroxene relics in their cores. The elongate orthopyroxene porphyroclasts define the foliation and in some cases contain clinopyroxene exolution lamellae. In some samples, orthopyroxene is highly deformed by kinking and in a few cases it is dynamically recrystallized with 0.1 mm neoblasts which are better preserved than the porphyroclasts from serpentinization and which do not contain clinopyroxene exolutions.

Clinopyroxene is not abundant in the peridotites and in some thin sections its modal content is very low. It forms porphyroclasts, which usually show polysynthetic mechanical twins and smaller recrystallized polygonal grains. In one sample, clinopyroxene consists of partially altered crystals

Table 1. Brief descriptions of selected samples collected during the Nautile dives.

Sample no.	*depth (m bsl)*	*Setting*	*Petrographic notes*
Dive HS 88-13			
13-02	3828	Talus	Serpentinized peridotite. One main set of serpentinite veins parallel to the foli-ation.
13-03	3778	Talus	Partially serpentinized peridotite. Two sets of serpentinite tension veins.
13-04	3724	Talus	Partially serpentinized peridotite with well-defined HT foliation. Two sets of cross-cutting serpentinite veins. Fractures filled by cal-cite cross-cutting all pre vious textures.
13-05	3728	Talus	Partially serpentinized peridotite cut by a cm-thick zircon-bearing gabbroic dikelet.
13-06	3726	Talus	Partially serpentinized peridotites with well defined HT foliation. One set of serpentine veins.
13-07	3590	Talus at foot of serpentine outcrop	Serpentinized peridotite with clinopy-roxene relics cut by fractures filled with amphibole, clay and chlorite.
13-08	3584	,,	Partially serpentinized peridotite. Two sets of serpentine veins cross-cutting each other at a high angle. Fractures filled by calcite.
13-09	3526	,,	Partially serpentinized peridotite. One set of serpentine veins.
13-10	3472	,,	Partially serpentinized peridotite cut by a cm-thick dikelet with altered pyrox ene (?). Talc along the dike-peridotite contact.
13-11	3229	,,	Partially serpentinized peridotite. One set of serpentine veins.
13-12	3100	,,	Partially serpentinized peridotite with mylonitic bands cross-cutting the HT foliation. Amphibole-bearing dikelet cross-cuts the rock parallel to the my lonites.
Dive HS 88-19			
19-01	3498	Talus	Breccia with serpentinized, spinel-bear ing dunitic clasts in contact with cm-scale, zircon-bearing chlorite.
Akademic Mtsislav Dredges			
DR1-2A			Breccia with serpentinized, spinel-bear-ing dunite clasts. Matrix consists of ser pentinite clasts and carbonate.

with concave boundaries which are interstitial in the serpentinite mesh texture (Fig. 3a). Interstitial clinopyroxene may result from melt impregnation, as described from Sites 395 (Boudier, 1979) and 670 (Cannat et al., 1990). Clinopyroxene is generally more resistant to alteration than orthopyroxene and olivine. Spinel forms red-brown coloured grains with irregular, sometimes "wormy" or holly-leaf shapes.

In the peridotite samples, a foliation is defined by the elongation of spinel and orthopyroxene porphyroclasts. This foliation (S_1) is produced by ductile deformation related to relatively high-temperature and low-stress conditions (Mercier and Nicolas, 1975; Nicolas and Poirier, 1976). In one sample (HS13-12A), S_1 is overprinted by mylonitic bands (S_2) in which olivine and pyroxene are recrystallized in new grains as small as 0.04 mm (Fig. 3b). This peridotite is cut by a 1 cm thick foliated vein. The internal foliation of the vein is marked by the elongation direction of brown amphibole crystals, which is almost parallel to the S_2 mylonitic foliation of the host peridotite (Fig. 3b). This suggests that the vein foliation and the mylonitic bands in the peridotite were produced by relatively high-stress plastic deformation at amphibolite facies metamorphic conditions. Other examples of high-temperature shear zones marked by the crystallization of synkinematic amphibole (hornblende) have been observed in peridotites from the same area (Casey, 1986; Gillis et al., 1993).

In the studied peridotites, serpentine is by far the most abundant alteration mineral (70%-100% serpentinization). Serpentinization of olivine accounts for the production of a typical "mesh" or "hour-glass" texture (Aumento and Loubat, 1971; Prichard, 1979; Wicks and Wittaker, 1977) outlined by magnetite dust. Alteration of orthopyroxene, that mostly produces brownish bastite, is concentrated along the porphyroclast rims and fractures. Exsolution lamellae of clinopyroxene may be preserved within bastite because clinopyroxene is more resistant to alteration than orthopyroxene.

The mesh texture in the samples is considered to have formed under static conditions because it is apparently undeformed. This texture is frequently cut by 0.1 to 0.5 mm thick cracks, filled with serpentine fibers which are thought to postdate the main serpentinization event in the samples. Two main sets of cracks have been observed: one set cuts the high temperature foliation at a small angle, the other cuts the foliation at a large angle. The relative timing inferred from cross-cutting relationships between these two sets of cracks does not yield systematic results. The serpentine fibers are usually undeformed, and are oriented perpendicular to the crack walls (Fig. 3c). The cracks may, therefore, be interpreted as extension veins which account for most of the volume increase during serpentinization. In hand specimen, this network of serpentine-filled cracks give a schistose appearance to the rock (see also Mével et al., 1991). Only in one specimen, 0.5 cm-sized veinlets are filled with sheared serpentine fibers. However, the abundant slickensided surfaces observed during the dives suggest that sheared

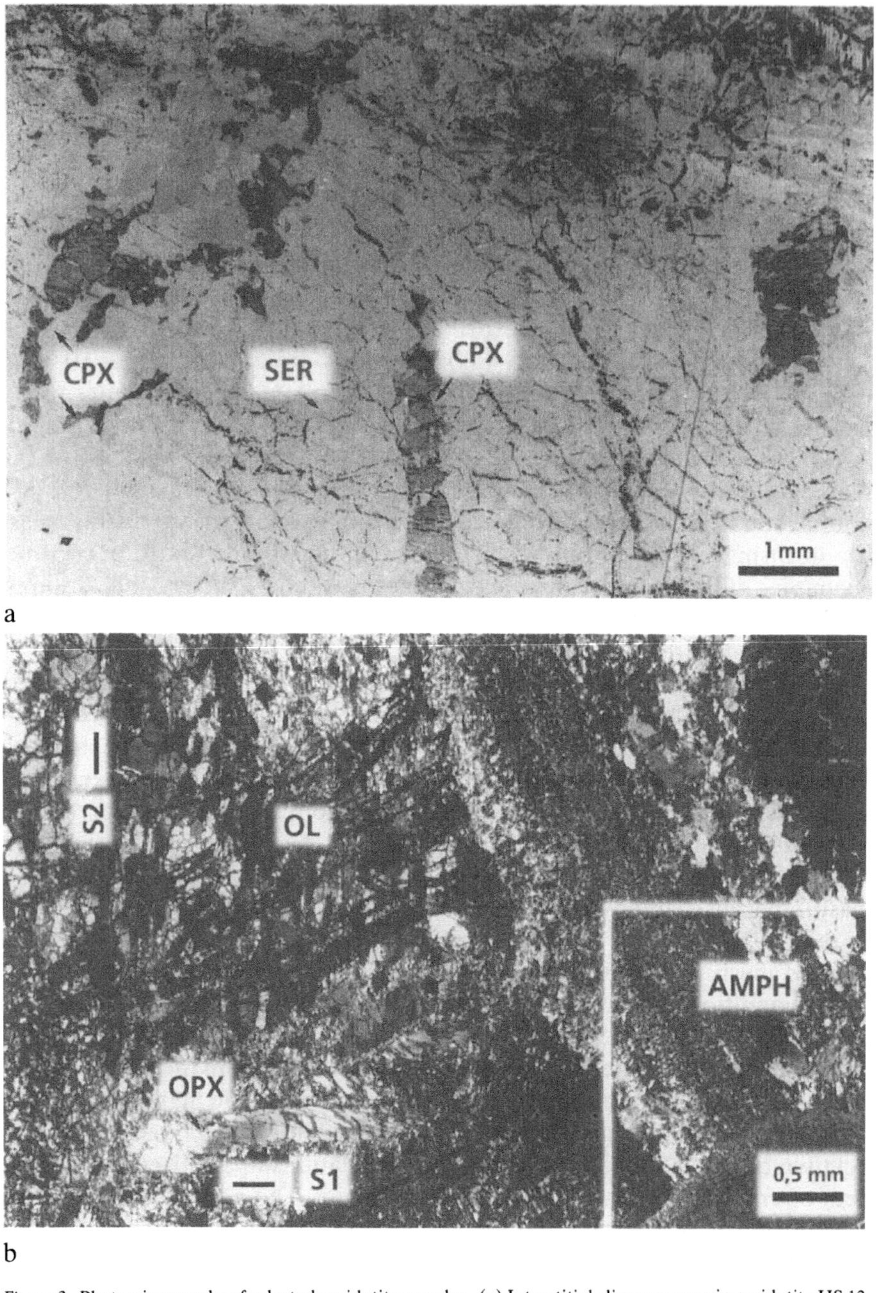

Figure 3. Photomicrographs of selected peridotite samples. (a) Interstitial clinopyroxene in peridotite HS 13-7 (plane polarized light): CPX, clinopyroxene; SER, serpentinized olivine. (b) Peridotite cut by foliated amphibole-bearing vein (sample HS13-12A, crossed nicols): S1, high-temperature foliation; S2, mylonitic foliation; OPX, orthopyroxene; OL, olivine (note orientation of OL olivine subgrains and neoblasts parallel to S2 foliation); AMPH, amphibole in foliated vein. Box corresponds to Fig. 3e.

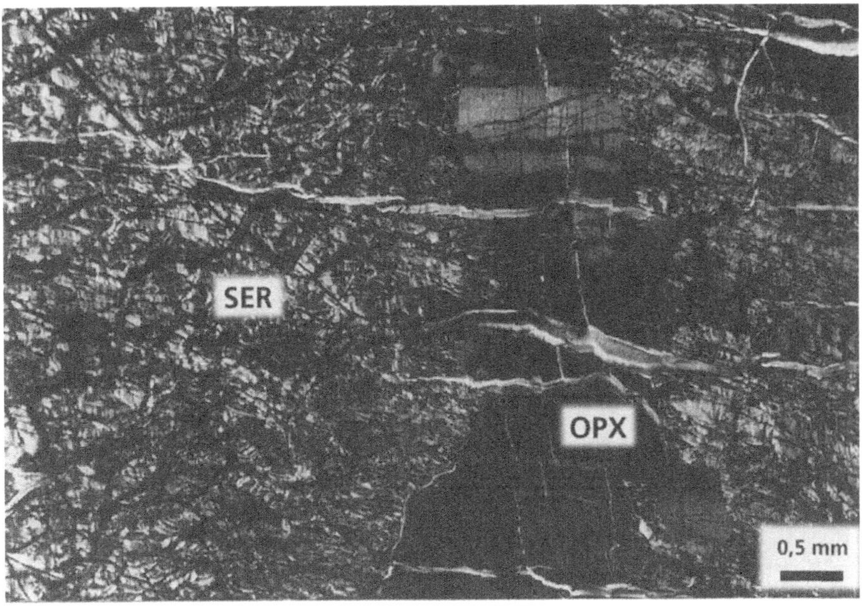

c

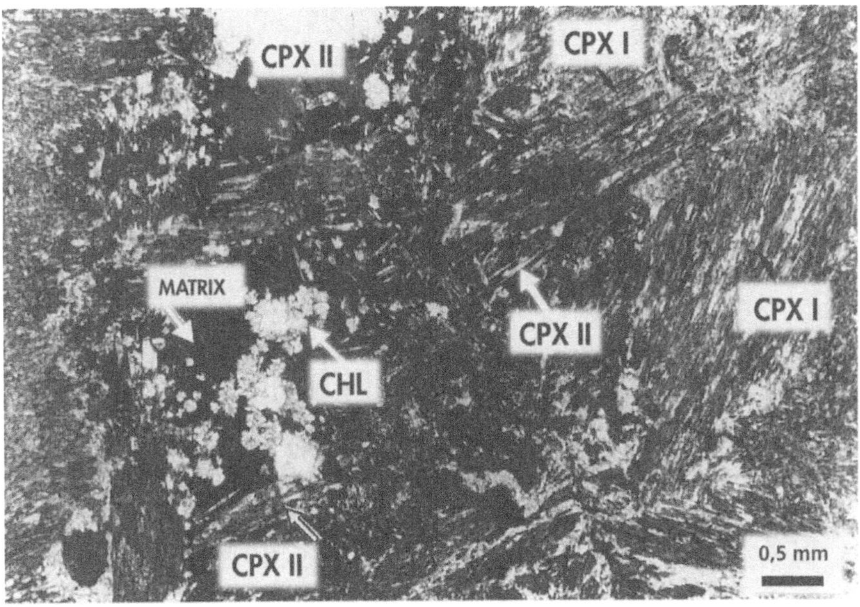

d

Figure 3 continued. Photomicrographs of selected peridotite samples. (c) Extensional veins filled with serpentine fibers cutting through the rock (crossed nicols). OPX, serpentinized orthopyroxene porphyroclast; SER, serpentinized olivine in groundmass. (d) Gabbroic dikelet, sample HS13-5 (plain polarized light). Contact is close to the left side; CPX I, clinopyroxene phenocrysts in dikelet; CPX II, differently sized Mn-rich clinopyroxene rimming CPX I or enclosed in matrix; CHL, Mn-chlorite in matrix.

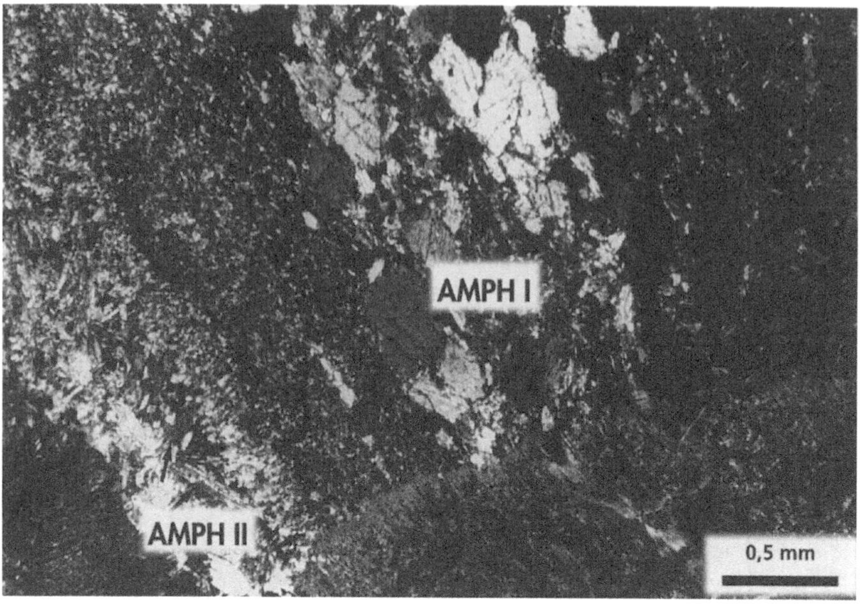

e

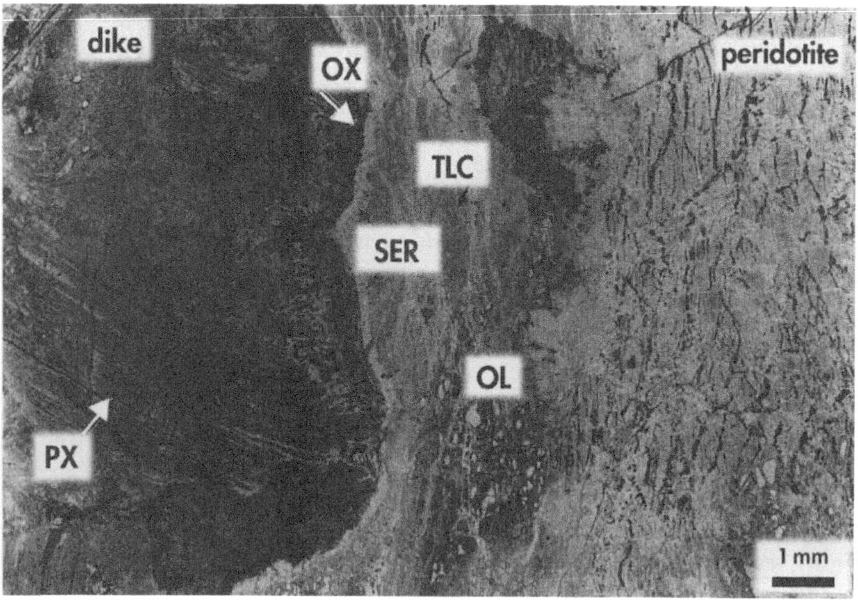

f

Figure 3 continued. Photomicrographs of selected peridotite samples. (e) Foliated amphibole-bearing vein, sample HS13-12A (see box Fig. 3b, crossed nicols); AMPH I, elongate brownish amphiboles defining internal foliation in vein; AMPH II, needle-shaped crystals of late amphibole rimming vein walls; note serpentine-filled vein at lower right corner, cutting amphibole-bearing vein and host peridotite. (f) Contact between peridotite and pyroxenite vein (sample HS13-10, plane polarized light); OL, olivine relics; TLC, talc after olivine near peridotite side; SER, serpentine rim along contact; OX, oxide + talc mixture near vein wall; PX, altered phenocrysts of probably pyroxene.

serpentine is more common than represented in our samples. Serpentine-filled veins are in places cut by millimetre wide fractures filled with carbonate, which therefore postdate the serpentine extension veins.

Hydrous alteration of the ultramafics also resulted in the crystallization of small amounts of amphibole which is usually found in fractures cross-cutting pyroxene crystals. Textural relationships between these amphibole veins and the serpentine mesh-texture or serpentine-filled cracks are not clear enough to establish a sequence of alteration events. Only in one case, a fracture filled with amphibole, clay and chlorite is cut by fibrous serpentine veins, clearly suggesting that serpentinization in this sample postdated the amphibole formation.

Textural characteristics of the veins cutting the serpentinized ultramafics

Four samples of serpentinized ultramafics (HS13-5, HS13-10, HS13-12A) contain millimetre to centimetre scale veins (Figs. 3b, 3d, 3e, 3f).

In sample HS13-5, the vein exhibits straight contacts with the surrounding peridotite and cuts the orthopyroxene and spinel foliation at a high angle. This vein contains zircon in clear grains, up to 2 mm in size, and prismatic centimetre-scale clinopyroxene crystals (CPX I in Fig. 3d) without a preferred orientation. This clinopyroxene is strongly altered into an unsolved brownish mineral and is fringed with limpid clinopyroxene crystals (CPX II in Fig. 3d). The groundmass of the vein is made up of a brownish microcrystalline material containing rectangular chlorite aggregates which look like pseudomorphs after plagioclase, fan-shaped chlorite flakes, and fine-grained colourless clinopyroxene crystals (CPX II in Fig. 3e). Amphibole crystals line the contact between the vein and the host peridotite. They are elongated orthogonal to the vein walls. This contact is in turn cut by serpentine-filled veinlets, demonstrating that serpentinization was, at least in part, later than the intrusion of the vein.

Although the veins are pervasively altered, their grainsize and texture, and also the occurrence of zircon and probable plagioclase pseudomorphs suggest a magmatic (gabbroic?) protolith. The centimetre-scale clinopyroxene crystals could be relics of the original (igneous) mineral assemblage, but they exhibit chemical compositions which are more consistent with a hydrothermal origin (see discussion below).

The peridotite HS13-10 is intruded by a centimetre-thick vein which exhibits irregular and sinuous walls (Fig. 3f). The vein cuts the peridotite foliation at a large angle. The primary mineral phases of the vein are no longer recognizable, but the coarse grain size and texture suggest a magmatic protolith. The vein essentially consists of curved centimetre-scale phenocrysts altered into serpentine, with fine opaque grains distributed along cleavages of the original crystals (Fig. 3f). The vein also contains millimetre-sized amphibole crystals which are bordered by opaque grains. These amphibole crystals and the serpentine pseudomorphs may derive from original pyrox-

Table 2. Selected microprobe analyses of olivine from the ultramafic samples. Structural formulae calculated on the basis of 4 oxygens. Analyses 1 and 2 refer to end points of analyzed traverse near peridotite-vein contact in sample HS12-5 (analysis 2 closest to contact). Analyses 3, 4, 5, 6: primary (I) olivine, dynamically recrystallized new grains (II) and olivine in the mylonitic bands (MYL) of sample HS 13-12A.

Sample	*HS13-12*	*HS13-12*	*HS13-12*	*HS13-10*	*HS13-5*	*HS13-5*
Analysis	*1 (I)*	*2 (II)*	*3 (MYL)*	*4 (I)*	*5 (I)*	*6 (I)*
SiO_2	40.92	40.8	42.57	40.85	37.52	36.2
TiO_2	0	0.05	0	0	0.02	0.01
Al_2O_3	0.03	0	0.03	0.01	0	0
Cr_2O_3	0.01	0	0	0.08	0	0.02
FeO	9.11	8.72	9.27	9.24	28.98	36.79
MnO	0	0.08	0.22	0.14	0.5	0.68
MgO	50.72	51.9	50.46	49.45	34.13	27.74
NiO	0.43	0.41	0.39	0.41	0.29	0.16
CaO	0.03	0	0	0	0.04	0.02
Na_2O	0.02	0.01	0	0.01	0	0.01
K_2O	0	0	0.02	0.01	0	0
Cl	0.02	0	0.01	0.02	0	0.01
Total	101.29	101.97	102.97	100.22	101.48	101.64
Si	0.989	0.979	1.009	0.998	0.995	0.997
Ti	0	0.001	0	0	0	0
Al	0.001	0	0.001	0	0	0
Cr	0	0	0	0.002	0	0
Fe^{2+}_{tot}	0.184	0.175	0.184	0.189	0.642	0.847
Mn	0	0.002	0.004	0.003	0.011	0.016
Mg	1.827	1.856	1.784	1.801	1.349	1.138
Ni	0.008	0.008	0.007	0.008	0.006	0.004
Ca	0.001	0	0	0	0.001	0.001
Na	0.001	0	0	0	0	0.001
K	0	0	0.001	0	0	0
Cl	0.001	0	0	0.001	0	0
Total	3.012	3.021	2.991	3.002	3.005	3.004
FO%	90.846	91.387	90.657	90.512	67.736	57.34
FA%	9.154	8.613	9.343	9.488	32.264	42.66

ene(s) of possibly different composition. This vein also contains spinel partially altered into magnetite. However, spinel grains have mostly been observed close to the contact with the peridotite, and may therefore be xenocrysts from the host rock. The groundmass of the vein consists of a mixture of serpentine, chlorite, and talc. Evidence for original plagioclase is lacking; this vein, therefore, may have originally been a pyroxenite. The vein-peridotite contact is marked by a millimetre-thick composite halo, formed by serpentine on the vein side and by talc replacing olivine on the peridotite side (Fig. 3f). This halo may represent a front of metasomatic reactions between the vein and the host peridotite. Timing relationships between serpentinization and metasomatic reactions are not clearly constrained because serpentinization of the peridotite is less extensive in the vicinity of the vein. However, extensional veins filled with serpentine fibers are cutting the peridotite-vein contact suggesting that extensive serpentinization took place after intrusion and metasomatic reactions.

Peridotite HS13-12 is cut by a centimetre-thick foliated vein (Figs. 3b, e).

The vein foliation lies parallel to mylonitic bands in the host peridotite and is marked by brown amphibole crystals and euhedral sphene. The brown amphibole occurs as mm-sized porphyroblasts and as dynamically recrystallized, elongate neoblasts parallel to the foliation. Needle-shaped, undeformed grains of colourless amphibole have grown near the vein walls in fan-shaped aggregates arranged with their elongation direction orthogonal to the walls (Fig. 3e). The matrix of the vein is composed of a mesh of serpentine, talc and opaques. The peridotite and the vein are in turn cut by cracks filled with stretched serpentine fibers (Fig. 3e). The metamorphic foliation of the vein prevents to recognize the original mineral assemblage. However, the occurrence of possibly magmatic relics in the cores of the amphibole porphyroblasts and the presence of sphene which may be derived from Fe-Ti oxides suggest that this vein could have a magmatic origin.

Finally, sample HS19-1 may be interpreted as a peridotite intruded by altered gabbroic veins. It consists of two parts: one part is made up of centimetre-sized, randomly oriented chlorite flakes containing limpid crystals of zircon. The other part is formed by 2-3 mm-sized clasts of serpentinized, spinel-bearing dunite embedded in a matrix of talc, chlorite and spinel. The first part may represent relics of a differentiated, zircon-bearing magmatic dikelet, comparable to that of sample HS13-5.

Mineral chemistry

About 450 microprobe analyses have been carried out on primary and secondary minerals in the sampled peridotites. Only selected analyses are reported in Tables 2 through 8, and these analyses may be considered as being representative of the obtained compositions. All analyses were performed using the automatized electron CAMECA-CAMEBAX microprobe of the Pierre et Marie Curie University of Paris and the CAMEBAX MBX electron microprobe of the IFREMER at Brest, France. The analytical conditions were: 10 kV current tension, 10 nA sample current, and 10 s/cycle counting time for all elements except for nickel and chromium (counting time of 15 s). A combination of oxydes and natural minerals has been utilized as standards.

Structural formulae of olivine, serpentine, chlorite, talc and clay have been calculated considering total Fe as FeO. Most of amphibole structural formulae show negligeable Fe^{3+} and were calculated considering total Fe as FeO. Only a few analyses of amphiboles yield appreciable Fe^{3+} contents. In this case, the Fe^{3+}-Fe^{2+} ratio was calculated by the method proposed by Vieten and Hamm (1978) and utilizing the following normalization, based on charge balance constraints: SUMCa: Ca / (Si + Al + Ti + Cr + Fe + Mn + Mg + Ca) = 15, if this sum of cations is > 15. Iron partitioning for spinel, orthopyroxene, and clinopyroxene have been obtained utilizing the MINTAB program for the Apple Macintosh computer (Rock and Carroll, 1990) based on the procedure proposed by Droop (1987).

Primary phases in the ultramafics

Primary phases of peridotites have been analyzed in order to compare their compostions with that of other serpentinized peridotites from axial valley outcrops.

The forsterite (Fo) content of olivine in the peridotites studied ranges from 89.4% to 91.3%, consistent with olivine compositions of many other oceanic, mantle-derived peridotites (Hamlyn and Bonatti, 1980; Bonatti and Michael, 1989; Dick, 1989). Selected olivine analyses are reported in Table 2 (analyses 1, 2, 3, 4). Recrystallization of olivine in the porphyroclastic texture and even in the mylonitic bands does not involve chemical variations (Table 2, analyses 2, 3). In sample HS13-5, however, the chemical composition of olivine is strongly modified in contact with the vein. In fact, olivine shows a Fo percentage ranging from 67.74% to 57.34% in a 1.37 cm long traverse across olivine crystals near the peridotite/dike contact (Table 2, analyses 5, 6). NiO values are slightly lower than away from contact. This Fe-enrichment in olivine suggests chemical exchanges between the peridotite and the intruding melt. Similar chemical exchanges have been described in the 15°37'-16°52'N peridotites cut by dikelets (Cannat et al., 1992). Such reactions imply a high temperature for the peridotites during intrusion, or a local reheating of the peridotite due to the intruding magma. Unlike sample HS13-5, the other studied peridotites cut by veins yield olivine compositions which are very homogeneous from the contacts towards the host rocks. This fact may be related either to a less contrasting composition of the intruding melt, or to lower temperature conditions during the intrusion event.

Selected analyses of orthopyroxene are reported in Table 3. The rims of the porphyroclasts are usually, but not systematically, depleted in Al and Cr with respect to the cores. Small recrystallized grains of orthopyroxene exhibit lower En and Al_2O_3 contents than the primary orthopyroxene (Table 3) probably due to recrystallization under lower temperature conditions (Sinton, 1979; Hamlyn and Bonatti, 1980). A comparison with other peridotites from the Mid-Atlantic Ridge shows that the analyzed orthopyroxenes have similar compositions, except for higher Mg/Mg+Fe values, as orthopyroxenes from peridotites of Sites 395 and 670, and from the North Atlantic (Michael and Bonatti, 1985).

Chemical differences between primary and recrystallized clinopyroxene are not as strong as in orthopyroxene (Table 4, analyses 1, 2, 3). Chemical composition of clinopyroxene exolutions inside orthopyroxene porphyroclasts are comparable to those of clinopyroxene porphyroclasts and of recrystallized grains (Table 4, analysis 4). The interstitial clinopyroxene in peridotite HS13-7 shows a composition comparable to that of the primary clinopyroxene in the other samples (Table 4, analysis 9).

Representative analyses of spinel are reported in Table 5 and have been compared with those of other published spinels (Fig. 4). The studied spinels have compositions quite similar to those of Site 395, and intermediate be-

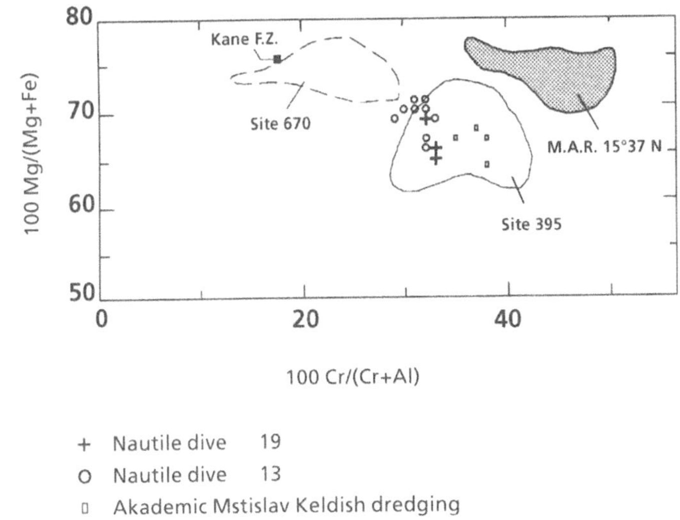

Figure 4. Diagram of the $Mg/Mg+Fe^{2+}$ vs. Cr/Cr+Al ratios in spinels of the studied samples. Spinel compositions from ODP Site 670, DSDP Site 395A, 15°37'N-16°52'N in the Atlantic (after Cannat et al., 1992) and from the Kane Fracture Zone (mean of 35 analyses, Mével, unpublished data) are shown for comparison.

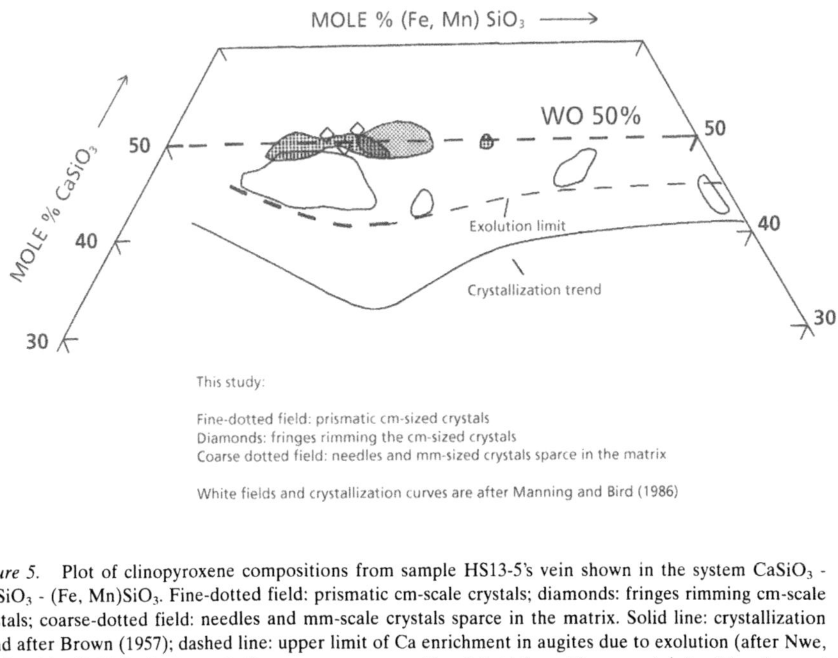

Figure 5. Plot of clinopyroxene compositions from sample HS13-5's vein shown in the system $CaSiO_3$ - $MgSiO_3$ - (Fe, Mn)SiO_3. Fine-dotted field: prismatic cm-scale crystals; diamonds: fringes rimming cm-scale crystals; coarse-dotted field: needles and mm-scale crystals sparce in the matrix. Solid line: crystallization trend after Brown (1957); dashed line: upper limit of Ca enrichment in augites due to exolution (after Nwe, 1976); white field: hydrothermal clinopyroxene compositional field from Skaergaard (after Manning and Bird, 1986).

Table 3. Selected microprobe analyses of primary (I) and dynamically recrystallized (II) orthopyroxene. C: core, R: rim. Structural formulae calculated on the basis of 6 oxygens.

Sample	*HS13-12*	*HS13-12*	*HS13-3*	*HS13-3*	*HS13-12*	*HS13-12*
Analysis	*1 (I)*	*2 (II)*	*3 (I)*	*4 (I)*	*5 (I,C)*	*6 (I,R)*
SiO_2	54.98	56.19	54.62	54.38	56.47	57.16
TiO_2	0.08	0	0.13	0.2	0	0.04
Al_2O_3	3.62	2.86	3.7	4.15	3.07	2.08
Cr_2O_3	0.79	0.76	0.91	0.88	0.83	0.62
FeO	5.54	5.42	6.4	5.91	6.23	6.02
MnO	0.18	0.19	0	0.3	0.09	0.04
MgO	33.87	34.46	32.91	33.04	33.5	33.99
NiO	0.17	0.06	0.27	0.11	0	0
CaO	1.38	0.52	1.94	1.26	1.87	1.76
Na_2O	0	0.01	0	0	0.05	0.01
K_2O	0	0.01	0	0.01	0	0.01
Total	100.62	100.48	100.9	100.25	102.11	101.72
Si	1.882	1.923	1.875	1.875	1.913	1.941
Al^{IV}	0.118	0.077	0.125	0.125	0.087	0.059
Al^{VI}	0.028	0.038	0.025	0.044	0.036	0.024
Ti	0.002	0	0.003	0.005	0	0.001
Cr	0.021	0.021	0.025	0.024	0.022	0.017
Fe^{3+}	0.063	0.019	0.068	0.048	0.032	0.017
Fe^{2+}	0.095	0.136	0.115	0.123	0.144	0.154
Mn	0.005	0.006	0	0.009	0.003	0.001
Mg	1.729	1.758	1.684	1.698	1.692	1.721
Ni	0.005	0.002	0.007	0.003	0	0
Ca	0.051	0.019	0.071	0.047	0.068	0.064
Na	0	0.001	0	0	0.003	0.001
K	0	0	0	0	0	0
Total	4	4	4	4	4	4
EN%	92.223	91.874	90.021	90.941	88.864	88.755

tween spinels of site 670 and of latitude 15°N-16°N. Spinel from peridotites dredged in the Kane Fracture Zone exhibits the least residual character with respect to the other plotted analyses (Mével, unpublished data).

In samples with a dunitic modal composition (DR1-2A, HS19-1; see Table 1), spinels have similar compositions suggesting no major difference in the degree of melting. In the diagram of Fig. 4 these spinels plot in the same area as spinels from Site 395, and very close to spinels of Nautile dive 13's peridotites.

Minerals of uncertain origin in the veins

Owing to the high extent of alteration which affected the veins, it is difficult to establish the precise origin of some minerals. For instance, the brownish amphibole filling the foliated dikelet in sample HS 13-12A may be regarded as a hydrothermal mineral that crystallized under ductile deformation conditions during fluid circulation. However, the cores of the amphibole porphy-

Table 4. Selected microprobe analyses of primary (I), dynamically recrystallized (II) and exsolution (EX) clinopyroxene from peridotites (analyses 1 to 4). Analyses 5 to 8 in vein sample HS 13-5: (5) cm-scale porphyrocryst; (6) secondary fringes at rim of porphyrocryst; (7) colourless secondary needles in vein matrix; (8) colourless mm-scale secondary crystal in matrix. Analysis 9: interstitial in peridotite. Structural formulae calculated on the basis of 6 oxygens.

Sample	*HS13-12A*	*HS13-5*	*HS13-5*	*HS13-7*	*HS13-5*	*HS13-5*	*HS13-5*	*HS13-5*	*HS13-7*
Analysis	*1(I)*	*2(I)*	*3(II)*	*4(EX)*	*5(I,V)*	*6(II,V)*	*7(II,V)*	*8(II,V)*	*9(I,INT)*
SiO_2	51.44	51.01	51.32	50.32	51.62	52.07	53.02	51.46	51.55
TiO_2	0.13	0.08	0.05	0.15	0.02	0.06	0.06	0.24	0.11
Al_2O_3	4.11	4.71	4.09	5.14	0.83	0.03	0	0.78	4.17
Cr_2O_3	1.34	1.23	1.34	1.28	0.1	0.13	0	0.01	1.01
FeO	1.95	2.44	3.57	2.27	11.55	10.56	6.02	9.96	2.25
MnO	0.03	0.03	0.05	0	0.68	1.77	2.31	1.06	0
MgO	16.46	17.09	16.12	16.57	10.23	10.68	13.18	11.27	16.72
NiO	0	0.02	0.12	0.07	0.03	0	0	0.12	0.06
CaO	23.74	22.5	22.73	23.67	24.38	24.27	25.01	24.89	23.41
Na_2O	0.24	0.39	0.35	0.12	0.03	0.01	0.02	0.02	0.12
K_2O	0	0.02	0	0.02	0	0.03	0	0	0.01
Total	99.44	99.52	99.74	99.61	99.47	99.61	99.62	99.81	99.41
Si	1.878	1.854	1.876	1.834	1.978	1.992	1.991	1.953	1.882
Al^{IV}	0.122	0.146	0.124	0.166	0.022	0.001	0	0.035	0.118
Al^{VI}	0.055	0.056	0.052	0.055	0.015	0	0	0	0.061
Ti	0.004	0.002	0.001	0.004	0.001	0.002	0.002	0.007	0.003
Cr	0.039	0.035	0.039	0.037	0.003	0.004	0	0	0.029
Fe^{3+}	0.037	0.074	0.055	0.069	0.004	0.01	0.017	0.047	0.03
Fe^{2+}	0.022	0	0.054	0	0.366	0.328	0.172	0.269	0.038
Mn	0.001	0.001	0.002	0	0.022	0.057	0.073	0.034	0
Mg	0.896	0.926	0.878	0.9	0.584	0.609	0.738	0.638	0.91
Ni	0	0.001	0.004	0.002	0.001	0	0	0.004	0.002
Ca	0.929	0.876	0.89	0.924	1.001	0.995	1.006	1.012	0.916
Na	0.017	0.027	0.025	0.008	0.002	0.001	0.001	0.001	0.008
K	0	0.001	0	0.001	0	0.001	0	0	0
Total	4	4	4	4	4	4	4	4	4
Mg/Mg+Fe	0.976	1	0.942	1	0.615	0.65	0.811	0.703	0.96
En%	48.513	51.382	48.197	49.342	29.951	31.528	38.503	33.225	48.818
Fs%	1.201	0	2.961	0	18.748	16.98	8.986	14.039	2.059
Wo%	50.287	48.618	48.843	50.658	51.301	51.492	52.511	52.736	49.123

Table 5 Selected microprobe analyses of spinel. C: core, R: rim. Structural formulae calculated on the basis of 24 cations.

Sample *Analysis*	*DR1-2A* *1 (C)*	*DR1-2A* *2 (C)*	*DR1-2A* *3 (C)*	*DR1-2A* *4 (R)*	*HS19-1* *5 (C)*	*HS19-1* *6 (R)*	*HS13-10* *7 (C)*
SiO_2	0	0.03	0	0	0.02	0.01	0.04
TiO_2	0.42	0.43	0.42	0.37	0.08	0.1	0.03
Al_2O_3	36.6	35.88	35.4	34.43	40.27	40.54	52.14
Cr_2O_3	29.6	31.49	31.69	31.6	28.87	28.14	17.04
FeO	17.61	16.28	16.89	18.11	14.33	14.49	7.72
MnO	0.29	0.22	0.05	0.26	0.16	0.08	0.13
MgO	15.93	15.96	15.82	14.81	16.57	16.69	22.83
NiO	0	0	0	0	0	0	0.05
CaO	0.02	0	0	0.02	0	0	0
Na_2O	0	0	0	0	0	0	0
K_2O	0	0	0	0	0	0	0
Total	100.76	100.29	100.27	99.6	100.33	100.05	99.98
Si	0	0.007	0	0	0.004	0.002	0.008
Ti	0.072	0.074	0.072	0.065	0.013	0.017	0.008
Al	9.804	9.654	9.551	9.427	10.627	10.699	11.192
Cr	5.319	5.684	5.735	5.804	5.111	4.982	4.728
Fe^{3+}	0.734	0.501	0.569	0.639	0.227	0.281	0.064
Fe^{2+}	2.614	2.607	2.664	2.879	2.457	2.432	2.368
Mn	0.056	0.043	0.01	0.051	0.03	0.015	0.04
Mg	5.397	5.432	5.399	5.129	5.531	5.571	5.584
Ni	0	0	0	0	0	0	0.016
Ca	0.005	0	0	0.005	0	0	0
Na	0	0	0	0	0	0	0
K	0	0	0	0	0	0	0
Total	24.001	24.002	24	24	24	24	24.008
Cr/Cr+Al	0.352	0.371	0.375	0.381	0.325	0.318	0.297
$Mg/Mg+Fe^{2+}$	0.674	0.676	0.67	0.64	0.692	0.696	0.702

roblasts exhibit Al and Ti contents comparable with those of magmatic or deuteric amphiboles (see below, and Fig. 7).

Another example is represented by clinopyroxene in the vein of sample HS13-5. It crystallized in three different textural sites, i.e., as strongly altered, centimetre-scale prismatic crystals without any preferred orientation, as smaller, clear grains fringing the rim of the centimetre-scale pyroxene crystals, and as colourless and limpid, millimetre-scale and smaller needle-shaped crystals sparsely arranged in the matrix (see Fig. 3d). Selected analyses of these clinopyroxenes are reported in Table 4 (analyses 5 through 8). When plotted in the pyroxene quadrilateral shown in Fig. 5, all points are scattered along the line "Wo50", with strong variations in Mg and (Fe+Mn) contents. A similar pattern was found in hydrothermal clinopyroxenes of the Skaergaard intrusion (Manning and Bird, 1986), in secondary clinopyroxenes in gabbros from ODP Site 735B (Stakes et al., 1991), and in hydrothermal clinopyroxenes in gabbronorites from the MARK area (Gillis et al., 1993). Similar to the hydrothermal clinopyroxene reported by Manning and Bird, the clinopyroxenes of this study have structural formulae showing higher Si and Ca values (cations p.f.u.) and lower Mg (cations p.f.u.). In the

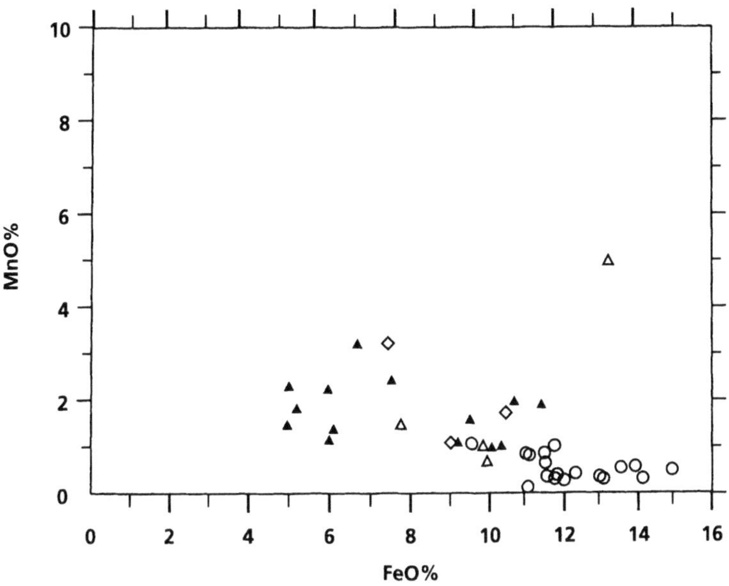

Figure 6. Diagram of MnO vs. FeO contents (wt%) in secondary clinopyroxene from sample HS13-5's vein. Open circles: cm-scale phenocrysts; open diamonds: crystal fringes at the rims of cm-scale phenocrysts; open triangles: mm-sized limpid crystals in matrix; filled triangles: limpid needles in matrix.

diagram of Fig. 5, all analyzed clinopyroxenes as well as the hydrothermal pyroxene from Skaergaard show Ca-enrichment with respect to both the calcic pyroxene crystallization trend as defined by Brown (1957) and Brown and Vincent (1963), and to the maximum Ca-enrichment limit for augite due to exsolution (Nwe, 1976). All clinopyroxenes in the vein considered may, therefore, be interpreted as secondary pyroxenes which crystallized from circulating hydrothermal fluid(s). In addition, these clinopyroxenes are compositionally distinct according to their different textural character. More specifically, the centimetre-scale prismatic crystals show Fe- and Al-enrichment and Mn-depletion with respect to the needle-shaped crystals sparcely arranged in the matrix (Table 4) which are strongly Mn-enriched (up to about 3% MnO). Pyroxene compositions from different textures in the vein have been plotted in the MnO-FeO correlation diagram of Fig. 6. MnO enrichment is gradually recorded from the centimetre-scale clinopyroxene to the smaller crystals in the matrix. This pattern may suggest that the veins were affected by circulation of hydrothermal fluid(s) with evolving compositions. The centimetre-scale clinopyroxene crystals may represent an early replacement of original magmatic augite. During circulation of hot and Mn-enriched fluids, these altered augites recrystallized along their rims. New crystals of Mn-rich clinopyroxene and Mn-rich chlorite (see Table 8) finally crystallized in the groundmass of the vein.

Table 6a. Selected microprobe analyses of amphiboles. Analyses 1, 2: core (C) and rim (R) of brown amphibole in foliated vein; 3, 4: core and rim of recrystallized amphibole neoblasts in vein; 5, 6: crystals with elongation direction orthogonal to the vein/peridotite contact in samples HS13-7 and HS13-12B. Structural formulae are calculated on the basis of 23 anhydrous oxygens.

Sample *Analysis*	*HS13-12A* *1(C)*	*HS13-12A* *2(R)*	*HS13-12A* *3(C)*	*HS13-12A* *4(R)*	*HS13-12A* *5(C)*	*HS13-12A* *6(R)*
SiO_2	50.70	52.04	51.80	51.91	45.34	58.29
TiO_2	2.18	1.64	1.54	1.74	1.24	0.01
Al_2O_3	5.93	4.52	4.56	5.24	11.01	0.26
Cr_2O_3	0.00	0.09	0.39	0.35	1.83	0.00
FeO	7.97	7.29	7.31	7.25	5.78	7.73
MnO	0.16	0.05	0.08	0.03	0.15	0.21
MgO	18.83	19.79	19.33	19.88	18.81	20.16
NiO	0.00	0.00	0.00	0.00	0.00	0.00
CaO	10.58	10.53	10.79	10.52	10.55	12.08
Na_2O	2.10	1.65	1.53	1.78	2.71	0.17
K_2O	0.17	0.11	0.12	0.10	0.17	0.03
Cl	0.00	0.00	0.12	0.00	0.00	0.00
TOTAL	98.62	97.71	97.57	98.80	97.59	98.94
Si	7.131	7.332	7.329	7.240	6.468	8.034
Al^{IV}	0.869	0.668	0.671	0.760	1.532	0.000
Sum T	8.000	8.000	8.000	8.000	8.000	8.034
Al^{VI}	0.114	0.083	0.089	0.102	0.320	0.042
Ti	0.231	0.174	0.164	0.183	0.133	0.001
Cr	0.000	0.010	0.044	0.039	0.206	0.000
Fe^{3+}	0.000	0.000	0.000	0.000	0.000	0.000
Fe^{2+}	0.707	0.577	0.627	0.544	0.341	0.815
Mn	0.000	0.000	0.000	0.000	0.000	0.000
Mg	3.948	4.156	4.076	4.133	4.000	4.142
Ni	0.000	0.000	0.000	0.000	0.000	0.000
Ca	0.000	0.000	0.000	0.000	0.000	0.000
Sum C	5.000	5.000	5.000	5.000	5.000	5.000
Mg	0.000	0.000	0.000	0.000	0.000	0.000
Fe^{2+}	0.231	0.283	0.239	0.302	0.349	0.076
Mn	0.019	0.006	0.010	0.004	0.018	0.025
Ca	1.595	1.590	1.636	1.572	1.613	1.784
Na	0.155	0.122	0.115	0.122	0.020	0.045
Sum B	2.000	2.000	2.000	2.000	2.000	1.930
Na	0.418	0.329	0.304	0.359	0.729	0.000
K	0.031	0.020	0.022	0.018	0.031	0.005
Sum A	0.449	0.349	0.326	0.377	0.760	0.005
$Mg/Mg+Fe^{2+}$	0.808	0.828	0.825	0.830	0.853	0.823

Alteration minerals.

Amphibole

Amphibole has been analyzed in peridotites as well as in veins (see Table 6 for representative compositions).

In peridotites, amphiboles filling fractures cutting pyroxene porphyroclasts (samples HS13-7, HS13-12B) range in composition from tremolite to Mg-hornblende or edenitic hornblende, and edenite (in Leake's 1978 classification). In sample HS13-7, tremolite is more abundant in the fracture center (see analysis 11, Table 6) whilst edenite and Mg-hornblende come from the

Table 6b Selected microprobe analyses of amphiboles. Analyses 7, 8: core (C) and rim (R) of crystals with elongation direction orthogonal to the vein/peridotite contact; 9, 10: pseudomorphs after probable pyroxene; 11, 12: crystals in fracture cutting peridotite in samples HS13-7 and HS13-12B. Structural formulae are calculated on the basis of 23 anhydrous oxygens.

Sample	*HS13-5*	*HS13-5*	*HS13-10*	*HS13-10*	*HS13-7*	*HS13-12B*
Analysis	*7(C)*	*8(R)*	*9(C)*	*10(R)*	*11(C)*	*12(C)*
SiO_2	57.14	56.48	55.48	56.63	55.42	48.57
TiO_2	0.00	0.00	0.07	0.05	0.06	0.37
Al_2O_3	0.93	1.50	2.33	1.86	2.19	10.06
Cr_2O_3	0.04	0.00	1.35	0.86	0.39	1.29
FeO	7.68	8.82	8.16	6.65	2.99	2.76
MnO	0.23	0.22	0.22	0.32	0.09	0.01
MgO	21.51	21.36	21.50	22.36	23.10	20.94
NiO	0.00	0.00	0.00	0.00	0.06	0.00
CaO	11.20	10.17	10.89	10.46	12.44	12.13
Na_2O	0.40	0.60	0.65	0.58	0.85	2.39
K_2O	0.03	0.01	0.00	0.00	0.02	0.00
Cl	0.01	0.00	0.01	0.02	0.03	0.01
TOTAL	99.17	99.16	100.66	99.79	97.63	98.50
Si	7.866	7.802	7.589	7.726	7.644	6.736
Al^{IV}	0.134	0.198	0.376	0.274	0.356	1.264
Sum T	8.000	8.000	7.964	8.000	8.000	8.000
Al^{VI}	0.017	0.046	0.000	0.025	0.000	0.382
Ti	0.000	0.000	0.007	0.005	0.006	0.038
Cr	0.004	0.000	0.146	0.093	0.043	0.141
Fe^{3+}	0.000	0.000	0.000	0.000	0.072	0.000
Fe^{2+}	0.564	0.556	0.463	0.330	0.123	0.111
Mn	0.000	0.000	0.000	0.000	0.000	0.000
Mg	4.414	4.398	4.383	4.547	4.750	4.328
Ni	0.000	0.000	0.000	0.000	0.007	0.000
Ca	0.000	0.000	0.000	0.000	0.000	0.000
Sum C	5.000	5.000	5.000	5.000	5.000	5.000
Mg	0.000	0.000	0.000	0.000	0.000	0.000
Fe^{2+}	0.321	0.464	0.471	0.430	0.151	0.208
Mn	0.027	0.026	0.025	0.037	0.011	0.001
Ca	1.652	1.505	1.596	1.529	1.839	1.802
Na	0.000	0.005	0.000	0.004	0.000	0.000
Sum B	2.000	2.000	2.093	2.000	2.001	2.011
Na	0.107	0.156	0.172	0.150	0.226	0.643
K	0.005	0.002	0.000	0.000	0.003	0.000
Sum A	0.112	0.158	0.172	0.150	0.229	0.643
$Mg/Mg+Fe^{2+}$	0.833	0.812	0.824	0.857	0.945	0.931

edges, close to the pyroxene porphyroclast. In sample HS13-12B, amphiboles mainly show compositions of edenite and edenitic hornblende (Table 6, analysis 12). They show higher Al contents than all the other amphiboles analyzed in fractures and in veins. Al^{IV} contents range between 0.31 and 1.37 (cations p.f.u.) and are always accompanied by a relatively high occupancy of the A site (Fig. 8). Unlike the Al-rich amphiboles filling the foliated vein, these fracture-filling amphiboles show relatively low Ti contents (Fig. 7). Similar compositions have been found in hydrothermal amphiboles replacing clinopyroxene from other peridotites of the MARK area (Gillis et al., 1993).

Table 7. Selected microprobe analyses of serpentine replacing olivine (OLV), orthopyroxene (OPX); in tension cracks (CR) and in vein (VEIN). Analyses 6, 7: in the matrix; analysis 8: pseudomorph after probable pyroxene. Structural formulae are calculated on the basis of 28 anhydrous oxygens.

Sample	*HS13-10*	*HS13-12*	*HS13-3*	*HS13-12*	*HS13-12*	*HS13-12A*	*HS13-12A*	*HS13-10*
Analysis	*1(OLV)*	*2(OLV)*	*3(OPX)*	*4(CR)*	*5(CR)*	*6(VEIN)*	*7(VEIN)*	*8(VEIN)*
SiO_2	43.97	43.56	40.06	43.78	45.04	44.61	42.79	42.14
TiO_2	0.06	0.01	0.00	0.00	0.00	0.09	0.03	0.01
Al_2O_3	0.09	0.33	3.06	0.77	0.39	0.83	1.57	6.09
Cr_2O_3	0.13	0.02	1.22	0.00	0.00	0.00	1.04	0.58
FeO	3.30	3.61	3.49	3.38	5.43	6.45	5.93	8.17
MnO	0.10	0.12	0.00	0.78	0.24	0.00	0.22	0.05
MgO	39.97	39.91	38.21	39.36	37.04	33.95	36.52	31.82
NiO	0.00	0.24	0.29	0.25	0.00	0.17	0.35	0.10
CaO	0.01	0.06	0.03	0.06	0.03	0.72	0.06	0.36
Na_2O	0.00	0.04	0.01	0.04	0.04	0.05	0.17	0.07
K_2O	0.02	0.00	0.02	0.09	0.02	0.04	0.09	0.00
Cl	0.04	0.01	0.08	0.03	0.07	0.02	0.00	0.03
Total	87.57	86.90	86.51	88.53	88.30	86.93	88.76	89.39
Si	8.162	8.095	7.619	8.093	8.359	8.450	7.999	7.834
Al	0.020	0.072	0.686	0.168	0.085	0.185	0.346	1.334
Fe^{2+}_{tot}	0.512	0.561	0.555	0.523	0.843	1.022	0.927	1.270
Mg	11.062	11.057	10.834	10.847	10.248	9.586	10.177	8.818
Ca	0.002	0.012	0.006	0.012	0.006	0.146	0.012	0.072
Na	0.000	0.014	0.004	0.014	0.014	0.018	0.062	0.025
K	0.005	0.000	0.005	0.021	0.005	0.010	0.021	0.000
Ti	0.008	0.001	0.000	0.000	0.000	0.013	0.004	0.001
Mn	0.016	0.019	0.000	0.122	0.038	0.000	0.035	0.008
Cl	0.013	0.003	0.026	0.009	0.022	0.006	0.000	0.009
Cr	0.019	0.003	0.183	0.000	0.000	0.000	0.154	0.085
Ni	0.000	0.036	0.044	0.037	0.000	0.026	0.053	0.015
Total	19.818	19.874	19.963	19.846	19.619	19.462	19.789	19.472
Mg/Mg+Fe	0.956	0.952	0.951	0.954	0.924	0.904	0.917	0.874

In veins, amphibole has been analyzed in the following different textural sites: (a) The brownish amphibole porphyroblasts in the foliated vein of sample

Table 8. Selected microprobe analyses of chlorite (1, 2), talc (3, 4) and mica (5, 6). FR: fracture cutting pyroxene porphyroclast in sample HS13-7; VEIN: magmatic veins in samples HS13-5 and HS13-12A; OLV: talc replacing olivine close to the peridotite/vein contact. Structural formulae of chlorite calculated on the basis of 28 anhydrous oxygens. Structural formulae of talc and mica calculated on the basis of 22 anhydrous oxygens.

Sample	*HS13-7*	*HS13-5*	*HS13-12A*	*HS13-10*	*HS13-7*	*HS13-12A*
Analysis	*1 (FR)*	*2(VEIN)*	*3(VEIN)*	*4(OLV)*	*5(FR)*	*6(VEIN)*
SiO_2	33.22	30.41	60.67	61.80	35.83	38.04
TiO_2	0.10	0.00	0.07	0.11	0.41	0.32
Al_2O_3	14.48	15.79	0.66	0.03	14.07	11.21
Cr_2O_3	0.99	0.00	0.00	0.01	0.25	1.32
FeO	5.81	15.40	2.06	2.34	6.18	6.54
MnO	0.09	0.92	0.16	0.00	0.11	0.04
MgO	31.86	24.83	30.60	29.54	27.18	28.86
NiO	0.18	0.01	0.04	0.49	0.23	0.15
CaO	0.03	0.05	0.03	0.04	0.21	0.04
Na_2O	0.00	0.00	0.31	0.08	0.70	0.11
K_2O	0.45	0.00	0.01	0.00	3.56	3.68
Cl	0.03	0.03	0.03	0.00	0.03	0.03
Total	87.23	87.43	94.62	94.44	88.75	90.33
Si	6.359	6.104	7.829	7.982	5.376	5.612
Ti	0.014	0.000	0.007	0.011	0.046	0.036
Al	3.267	3.735	0.100	0.005	2.488	1.949
Cr	0.150	0.000	0.000	0.001	0.030	0.154
Fe^{2+}_{tot}	0.930	2.585	0.222	0.253	0.776	0.807
Mn	0.015	0.156	0.017	0.000	0.014	0.005
Mg	9.091	7.430	5.887	5.688	6.080	6.348
Ni	0.028	0.002	0.004	0.051	0.028	0.018
Ca	0.006	0.011	0.004	0.006	0.034	0.006
Na	0.000	0.000	0.078	0.020	0.204	0.031
K	0.110	0.000	0.002	0.000	0.682	0.693
Cl	0.010	0.010	0.007	0.000	0.008	0.008
Total	19.979	20.033	14.157	14.015	15.765	15.666
Mg/Mg+Fe	0.907	0.742	0.964	0.957	0.887	0.887

HS13-12 exhibit compositions ranging from Mg-hornblende in the core to Mg-hornblende/actinolite in the rim (see Table 6, analyses 1 and 2, respectively). The recrystallized neoblasts in the foliation of the same vein show Mg-hornblende and actinolite compositions which may be found simultaneously in the same textural site (core and rim), and do not reflect any systematic zonation. Both amphibole porphyroblasts and their finer grained neoblasts have relatively high Ti contents. In the Cr vs. Ti diagram shown in Fig. 7, their compositions (open and filled diamonds) plot in an isolated field characterized by high TiO% and low Cr_2O_3% values. Notably, the porphyroblast cores exhibit higher Ti contents (> 2.0%) than all other amphiboles analyzed (Table 6 and Fig. 7). All amphiboles have relatively high Al^{IV} (Table 6). Moreover, increasing Al^{IV} contents are usually accompanied by increasing site A occupancy (Fig. 8). The Al-Ti-rich porphyroblast cores are inferred here to be relics of magmatic amphiboles that probably re-equilibrated during hydrothermal alteration.

(b) The amphibole crystals in the pyroxenite vein of sample HS13-10 exhibit

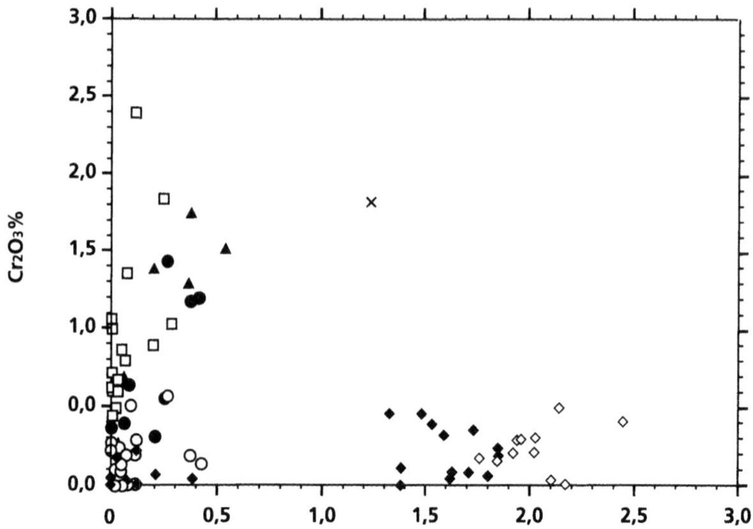

Figure 7. Diagram of the Cr_2O_3 vs. TiO_2 contents (wt%) in the studied amphiboles. Filled circles: peridotite fracture cutting pyroxene in sample HS13-7 (see text); filled triangles: fracture cutting pyroxene in sample HS13-12B; open circles and cross: (all samples) along vein/peridotite contact (cross corresponding to pargasitic-hornblende of sample HS 13-12A, see text and Table 6); open squares: pyroxenite vein of sample HS13-10; diamonds: foliated vein of sample HS13-12A (open: core of the porphyroblasts; filled: rim of the porphyroblasts and neoblasts).

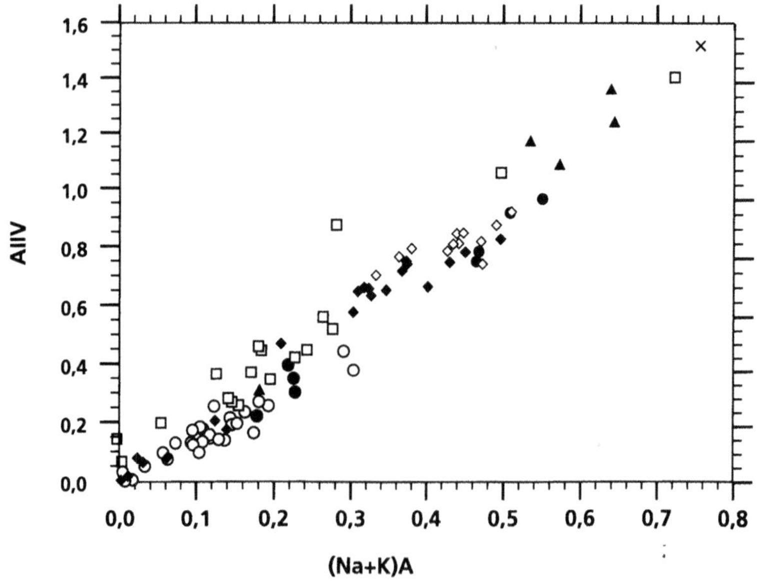

Figure 8. Diagram of the Al^{IV} vs. (Na+K)A contents (cations p.f.u.) in the studied amphiboles. Same symbols as Fig. 7.

Mg-hornblende and actinolite compositions. These amphiboles are also characterized by high Cr_2O_3 contents (between 0.10% and 2.40 Cr_2O_3%, see also Fig. 7). On the basis of this characteristic we suggest that these amphiboles represent pseudomorphic replacements of original chromium-bearing pyroxenes.
(c) The amphibole crystals rimming the vein walls are characterized by comparable chemical compositions in the three studied samples. They exhibit a chemical zonation ranging from Mg-cummingtonite in the core to Mg-hornblende and actinolite at the rim. In the foliated vein, the zonation ranges between pargasitic-hornblende in the core, characterized by Ti contents as high as about 1% TiO_2 (see analyses 5, Table 6 and Fig. 7) and actinolite in the rim (analysis 6). All amphiboles analyzed near the walls of the foliated vein of sample HS 13-12A are characterized by slightly higher TiO_2 and Cr_2O_3 contents and lower FeO contents than amphiboles in other samples. This may reflect the overall higher Ti content of the vein. Amphibole from sample HS13-5 vein is characterized by high FeO content, FeO% ranging around 9-12%. These values probably reflect high iron contents of the intruding magma, which also affected the composition of olivine in the host peridotite as discussed above.

Chemical compositions obtained from the amphiboles analyzed in peridotites and veins are well arranged in a pargasitic trend (Al^{IV} vs. (Na+K)A diagram, Fig. 8).

Serpentine

X-ray-diffraction data on serpentine from different textural domains suggest that chrysotile is the most widespread type. It mainly occurs in the mesh texture after olivine, where it is associated with minor amounts of antigorite. In the serpentine veins, chrysotile is present with minor lizardite. In one sample (HS 13-12), lizardite is the main component of the serpentine veins.

Chemical analyses of serpentine replacing olivine, orthopyroxene and clinopyroxene in the peridotite samples are listed in Table 7. Serpentine from these different textural sites shows very similar compositions, except for serpentine replacing orthopyroxene that show anomalous high Al_2O_3 and Cr_2O_3 contents (see analyses 3 and 8, Table 7). This feature may have been inherited from the orthopyroxene composition, suggesting that Cr and Al may be considered as relatively immobile elements during the serpentinization of ultramafic rocks (Dungan, 1979; Agrinier et al., 1988).

In veins, serpentine occurs as a mineral component of the matrix and in pseudomorphic replacements. It generally shows higher FeO contents than serpentine analyzed in the peridotites (Table 7) and it is usually very poor in Cr_2O_3. Only in the foliated vein of sample HS13-12, serpentine locally shows high Cr contents (up to 1.04% Cr_2O_3, see analysis 7, Table 7). In the pyroxenite vein of sample HS13-10, serpentine replacing deformed centimetre-scale phenocrysts has relatively high Cr_2O_3 content (0.58 Cr_2O_3%, analysis 8, Table 7) consistent with our interpretation that the original mineral was a pyroxene.

Chlorite
Chlorite occurs in the vein matrix and in peridotites where it has been detected replacing pyroxene along with serpentine, and together with amphibole and clay as filling mineral in fractures cutting pyroxene porphyroclasts. Chemical compositions of chlorite from these different textural sites are very similar. The only exception is represented by chlorite which forms fan-shaped aggregates in the matrix of sample HS13-5's vein. Here, chlorite shows relatively high MnO content (0.11% - 0.92% MnO) when compared to the other samples (Table 8). The Mn-enrichment of chlorite seems to be consistent with the Mn-enrichment of secondary clinopyroxene in the same sample (see Table 4).

Talc
Talc occurs in the matrix of the foliated vein of sample HS13-12, and near the contact between the pyroxenite vein and the host peridotite in sample HS13-10. In this case, talc is present either on the peridotite side where it replaces olivine, or along the vein walls in the interstitial matrix. Representative analyses are listed in Table 8. Chemical compositions of talc at these different textural sites are the same. However, the NiO content of talc replacing olivine (sample HS13-10) is higher than in other sites (see Table 8, analysis 4), suggesting the possibility that talc may inherit the composition of olivine.

Clay minerals
Clay minerals have been analyzed in a fracture cutting a pyroxene porphyroclast in peridotite HS13-7, where it is associated with chlorite, amphibole and serpentine. The chemical composition of this mineral (Table 8) suggests that it is made up of interlayered vermiculite (Mg, Ca, Al contents) and illite (K content).

Discussion

In the following section we discuss the petrological features of the MARK peridotites and included veins, and use the mineral assemblages, their compositions and textural features to constrain the nature of the veins. We then address the relative timing of peridotite emplacement and vein intrusion, and discuss how these events fit within tectonic models proposed so far.

Primary characteristics of the peridotites

The petrographic and chemical characteristics of the peridotites suggest that they are mostly harzburgitic, mantle-derived ultramafics. The complete serpentinization of olivine in our dunitic samples precludes a straightforward textural interpretation, however, on the basis of their spinel composition the dunites may be interpreted as mantle-derived rocks rather than as magmatic cumulates, the latter being usually characterized by strongly Cr- and Fe-enriched spinels as compared with mantle tectonites (Hébert et al., 1989, and references therein). The composition of spinel in the dunites is comparable to that of spinel in tectonitic harzburgites from the MARK area but, consistent with a more depleted character than harzburgites, it is characterized by higher Cr and Fe contents.

The composition of olivine (Fo contents between 89.4% and 91.9%) in the harzburgites is comparable to that of other oceanic mantle-derived peridotites (Hamlyn and Bonatti, 1980; Bonatti and Michael, 1989; Dick, 1989), including the other ultramafic rocks from the MARK area (DSDP Site 395, ODP Site 670). Chemical compositions of orthopyroxene and spinel suggest, however, that the rocks studied here have a slightly more depleted character than the peridotites drilled in Site 670 and that they are quite comparable to the peridotites drilled in DSDP Site 395. These observations support the existence of local variations in the degree of mantle partial melting in peridotites, over distances of a few tens of kilometres or less, as discussed for peridotites of the Mid-Atlantic Ridge by Bonatti et al. (1992, and references therein). The amplitude of these local variations remains moderate, however, when compared with longer-wavelength variations such as those detected between the MARK area peridotites and the significantly more depleted samples from 15°-16°N in the Atlantic (Cannat et al., 1992).

Nature of the veins and timing of intrusion

In spite of alteration which prevents the analyses of the original mineral assemblage of the veins, the occurrence of some mineral relics and the chemistry of the alteration phases allow us to make some inferences on the nature of the veins.

(a) The concurrent presence, in the foliated vein, of relict magmatic Ti-rich hornblende and of sphene as probable alteration product of Fe-Ti oxide, and also the general high Ti content of amphiboles, suggest that this vein probably crystallized from igneous fluids with Ti-(Fe?)-rich compositions (evolved basaltic magma?) that intruded the peridotites. The magmatic assemblage likely re-equilibrated during hydrothermal alteration as suggested by evolving amphibole compositions.

(b) The Fe-rich composition of amphiboles in the zircon-clinopyroxene-bearing vein, and also the occurrence of plagioclase phantoms and zircon are in agreement with the interpretation of this vein being a minor intrusion of evolved magmatic fluid (Fe-rich gabbro?). The composition of the fluid may account for the chemical variations in olivines of the host peridotite, and it may also explain the higher Fe content of the centimetre-scale clinopyroxene crystals of the vein as compared with the fine-grained pyroxene in the interstitial matrix. The bigger ones may be interpreted as relics of magmatic pyroxene which were successively affected by hydrothermal fluid circulation.

(c) The occurrence of Cr-rich amphibole and Cr-rich serpentine probably replacing original pyroxene(s) in a plagioclase-free vein supports the presence of pyroxenite dikelets. The fact that the original pyroxene was replaced by secondary minerals of different compositions, namely a Ca-bearing and a Ca-free mineral, suggests that there were probably two pyroxene types, i.e. clinopyroxene and orthopyroxene. Whether spinel pertains to the mineral assemblage of the dikelet or not is uncertain, because its textural site is not well constrained. The nature of the interstitial matrix, which consists of serpentine, chlorite and talc suggests that it was originally composed by Mg-Fe-bearing minerals (olivine?). One sample of websterite has been drilled in ODP Site 670 (Detrick, Honnorez et al., 1988), where it is surrounded by serpentinized peridotites. This websterite contains clinopyroxene and orthopyroxene in almost equal amounts, with interstitial olivine and Cr-spinel (Juteau et al., 1990). Plagioclase-free, olivine-websterites have also been described from the Kane and Islas Orcadas Fracture Zones, where they occur as bands in residual peridotites (Dick et al., 1984). These websterites have been interpreted as "in situ" crystallized trapped melt.
(d) The high degree of alteration in the zircon-chlorite-bearing vein prevents any certain interpretation, although it might be comparable to the zircon-clinopyroxene-bearing dikelet.

On the basis of the mineral assemblages, their composition, and also textural features, we argue that the veins cutting the peridotites, collected during Nautile dives 13 and 19, are the product of crystallization from igneous melts characterized by a wide range of compositions. The timing of intrusion of these melts may be inferred by considering the textural relationships between the host peridotites and the dikelets. We may argue that the igneous melts intruded the harzburgites at some stage after the low-stress, high-temperature ductile deformation under mantle conditions producing the high-temperature foliation in the ultramafics. In one case we can infer that magma intrusion occurred before completion of ductile deformation under higher stress, lower temperature conditions involved in the development of the mylonitic shear zones in the peridotite and the dynamic recrystallization of hornblende in the dikelet. Because olivine did behave plastically and recrystallized during deformation, it can be inferred that this occurred at temperatures above the olivine brittle-ductile transition (experimentally found around 700°C - 1000°C depending on the strain rate, Kirby, 1983), which is partly consistent with the hornblende stability field (Spear, 1981). In all other cases studied here, dikelets do not show any evidence of ductile deformation. The occurrence of both foliated and undeformed dikelets cutting the tectonitic harzburgites may suggest that magmas intruded the peridotites while affected by inhomogeneous deformation (e.g., shear zones and undeformed domains). On the basis of the present observations only, however, it is difficult to constrain the depth at which intrusions occurred, and even the relative timing among different intrusion events. Finally, textural features suggest that melt intrusion occurred before complete serpentinization of the ultramafic rocks, causing metasomatic reactions in places between the host rock and the intruding dikelets. This means that the host rock was likely still hot at the time of intrusion or, alternatively, that the injection of magma after serpentinization caused some reheating of the peridotites.

Conditions of hydrothermal alteration

The peridotites and dikelets are extensively altered. Petrographic and microchemical features of the secondary mineral assemblages suggest that hydrous alteration attained temperature conditions compatible with amphibolite facies metamorphism. This is attested by the occurrence of Mg-hornblende, pargasitic-hornblende and of coexisting hornblende-actinolite pairs which suggest temperatures of formation in the range of 450° to 600°C as discussed by Robinson et al. (1981) for the calcic amphibole solvus diagram. Only in one case we may assert that high-temperature alteration was accompanied by the development of a metamorphic foliation. This suggests that fluids could have penetrated along localized zones of ductile deformation. Differently, in most cases, the hydrous alteration occurred in the brittle rheological regime. Lower temperature conditions for the hydrous alteration affecting the studied samples are mostly represented by the assemblage of actinolite, chlorite, sphene and chrysotile. If we take into account, as is generally agreed, that Al^{IV} content in amphibole is dominated by temperature (Leake, 1965; Liou et al., 1974; Blundy and Holland, 1990), these generally retrogressive metamorphic conditions are also emphasized by the Al^{IV} vs. (Na+K)A diagram, in which the analyzed amphiboles plot on a pargasitic trend. On the other hand, this compositional trend must be taken with caution. In fact, among all the analyzed amphiboles, only amphibole crystals occurring along the dikelets walls are characterized by zonation ranging from pargasitichornblende/hornblende to actinolite, and may therefore account for a progressive reequilibration under decreasing temperatures.

The studied rocks were also affected by the circulation of Mn-enriched hydrothermal fluids which account for the crystallization of secondary Mn-rich clinopyroxene and chlorite, and probably for the alteration of the primary (magmatic) pyroxene in one dikelet. Secondary clinopyroxene is characterized by different MnO%, attesting that it crystallized from fluids with evolving composition. The circulation of these hydrothermal fluids seems to be restricted to dikelets, as suggested by the fact that no secondary minerals with similar Mn-enriched composition have been observed in the host peridotites. On the other hand, the entire hydrous alteration which affected the studied rocks is concentrated along dikelets and fractures. This affirms that dikelets and fractures represented the preferential pathways for hydrothermal fluid circulation. Finally, textural observations suggest that such fluid circulation occurred before the last stage of serpentinization allied with serpentine extensional veins cutting the peridotite-dikelet contacts.

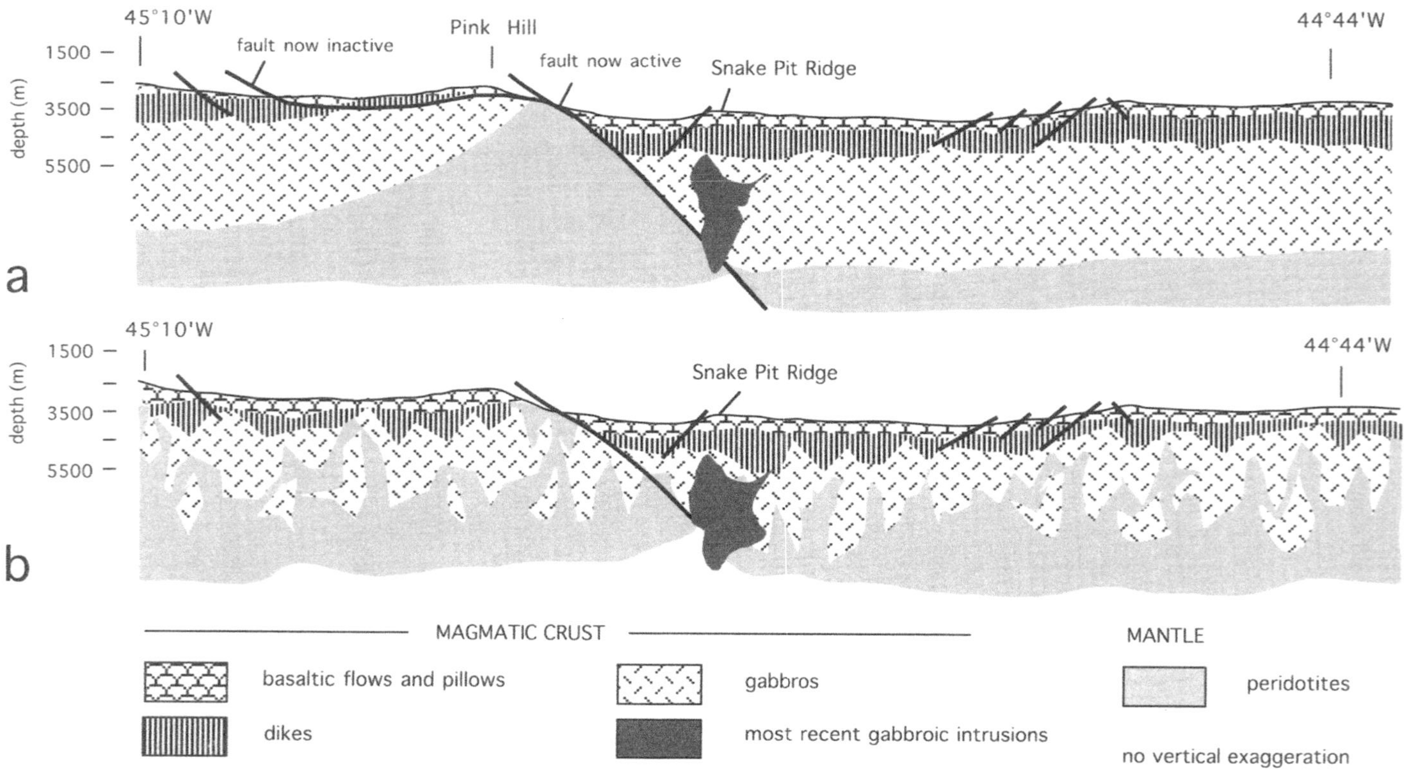

Figure 9 Interpretative geological sections across the Pink Hill area, according to the models explained in the text: (a) low-magma budget spreading following a high-magma budget period; (b) consistently low magma supply. Net displacement along the western median-valley wall master fault is about 6000 m in the first case and about 1000 m in the second case.

Serpentine crystallized in a wide range of structural and textural situations. This suggests that serpentinization may have occurred during different stages of peridotite emplacement. Furthermore, the occurrence of different serpentine polymorphs and namely of antigorite in the mesh texture supports the hypothesis that hydrothermal alteration may have affected the peridotites and the included dikelets under different temperature conditions.

Gabbroic dikelets in the MARK peridotites as minor intrusions in oceanic lithosphere created at magma-starved spreading centres.

Three possible mechanisms leading to the emplacement of upper mantle and lower crustal rocks in the seafloor have been so far envisaged: intense stretching of a normal-thickness crust along ductile and brittle normal shear zones (Karson, 1991), extension of a thin and heterogeneous crust produced during magma-poor spreading (Mével et al., 1991; Cannat, 1993), and serpentinite diapirism (Francis, 1981). These mechanisms may be expected to produce distinct crustal structures and crust-mantle relationships. In the case of intense stretching of a normal-thickness crust (Fig. 9a) one would expect a layered, albeit tectonically dismembered crust and crust-mantle transition zone, comparable to those seen in some ophiolites (e.g., the Josephine ophiolite; Harper, 1984). In the case of serpentinite diapirism one may predict the development of a melange of gabbro, basalt and peridotite blocks in a serpentine matrix, comparable with recent findings in the Mariana forearc (Fryer, Pearce et al., 1990). The MARK area peridotites and gabbros form coherent outcrops, with ridge-parallel, and locally transform-parallel normal faults (Karson and Dick, 1983; Mével et al., 1991), and they exhibit none of the structural and lithological complexities to be expected in the serpentinite diapirism case. In addition, the peridotites contain mylonitic shear zones which are likely related to the extension of the lithosphere, and which developed under amphibolite facies conditions, i.e., before the onset of serpentinization (Casey, 1986; this study). Finally, the MARK area peridotite and gabbro outcrops do not coincide with gravity lows (Bergès, 1989; Morris and Detrick, 1991), as would be expected if they were underlain by low density serpentinites. We therefore do not favour a serpentinite diapirism mechanism for peridotite and gabbro emplacement in this part of the Mid-Atlantic Ridge. A model involving intense stretching of a normal-thickness crust (Fig. 9a) is more consistent with the submersible observations of low-angle normal faults in the peridotite and gabbro outcrops (Karson and Dick, 1983; Mével et al., 1991), and with geophysical data indicating thin crust (4 km or less; Cormier et al., 1984; Purdy and Detrick, 1986; Morris and Detrick, 1991) beneath the median valley in the 23°18' N to 23°34' N area. According to this model, a continuous and normal-thickness oceanic crust is produced during a period of high magma budget, which is in turn followed by a period of low magma budget. During this magmatically starved episode, spreading is accomodated by lithospheric stretching, and mantle rocks can be emplaced tectonically. The model does not, however, fit the results of the sample studies, which we argue are more compatible with a model involving a thin and heterogeneous crust, essentially characterized by low magma supply (Fig. 9b). These results concern: (1) the varied, and locally very fractionated chemistry of gabbroic melts intrusive into the peridotites, (2) the relative timing of deformational and intrusive events in the peridotites, (3) the primary mineralogy of the peridotites, and more specifically the lack of troctolites, wehrlites and other rock types typical of the transition zone in many ophiolites (Nicolas and Prinzhofer, 1983). In a layered, normal thickness crustal model (relatively fast-spreading rate and/or high heat supply from the asthenosphere), the uppermost mantle peridotites may be intruded by gabbros, as frequently observed in ophiolites (Nicolas and Jackson, 1982; Nicolas et al., 1988). According to that same model, however, these gabbros are not expected to have evolved compositions (Sleep, 1975), but should represent primitive melts extracted from the underlying mantle. The evolved (i.e., Fe-, Ti- and Zr-rich) melt compositions inferred for our MARK area dikelets are more likely to represent end products of extensive fractionation in nearby magma bodies. The timing of these gabbroic intrusions with respect to the deformational history of the peridotites is also inconsistent with a normal-thickness crustal model (ridges with a long-lasting magma chamber at crustal depths and relatively thin lithosphere, Cannat, 1993) in which the uppermost mantle at the ridge axis should deform at high temperatures close to the solidus and low deviatoric stresses (Nicolas et al., 1988; Ceuleneer et al., 1988). Gabbroic intrusions in most of our MARK area samples clearly postdate the end of ductile flow, and were locally contemporaneous with lower temperature (amphibolite facies), higher stress (lithospheric?) deformation of their host peridotite. This fits better with a thin and heterogeneous crustal model (Fig. 9b), in which gabbros should be emplaced into mantle rocks already uplifted and cooled into the axial lithosphere (Cannat et Casey, this volume). Similar highly fractionated gabbroic intrusions into mantle-derived peridotites have been found so far along the Mid-Atlantic Ridge (15°37'N, Cannat et al., 1992; Cannat and Casey, this volume) and along the Southwest Indian Ridge fracture zones (Fisher et al., 1986; Bloomer et al., 1989; Engel and Fisher, 1975), that is, in tectonic settings characterized by the occurrence of geophysically thin crustal sections. Finally, the lack of wehrlitic and troctolitic rocks, and the scarcity of dunites in our sample set indicates that the MARK area ultramafic outcrops are not comparable with the transition zone, found at the crust-mantle boundary in many ophiolites and attributed to high-temperature interactions between basaltic melts, upper mantle rocks and early gabbroic and ultramafic cumulates at the base of the axial magma chamber (Nicolas and Prinzhofer, 1983; Karson et al., 1984; Benn et al., 1988).

Acknowledgements

We would like to thank J. Karson and R. Caby for careful reviews which improved the manuscript. We also thank technicians of the U.P.M.C. of Paris and IFREMER of Brest for their assistance with the microprobe analyses. Financial supports to P. Tartarotti from the University of Paris 7 (N. 372 MCF 0697) and from a C.N.R. fellowship of Italy are gratefully aknowledged.

References

Agrinier, P., Mével, C. and Girardeau, J., 1988. Hydrothermal alteration of the peridotites cored at the ocean/continent boundary of the Iberian margin: petrologic and stable isotope evidence. In: G. Boillot, E.L. Winterer et al. (Editors), Proc. of the ODP, Scientific Results, 103: 225-234.

Aumento, F. and Loubat, H., 1971. The Mid-Atlantic Ridge near 45°N. XVI. Serpentinized ultramafic intrusions. Canad. Journ. Earth Sci., 8: 631-663.

Benn, K., Nicolas, A. and Reuber, I., 1988. Mantle-crust transition zone and origin of wehrlitic magmas: evidence from the Oman ophiolite. Tectonophysics, 151: 75-85.

Bergès, J.P., 1989. Modélisation gravimétrique de la ride Médio-Atlantique: mur Ouest et Snake Pit à la latitude 23°20'N, Rapport DEA, Université de Paris VI.

Bloomer, S.H., Natland, J.H. and Fisher, R.L., 1989. Mineral relationships in gabbroic rocks from fracture zones of Indian Ocean ridges: evidence for extensive fractionation, parental diversity, and boundary-layer recrystallization. In: A.D. Saunders and M.J. Norry (Editors), Magmatism in the Ocean Basins. Geol. Soc. London Spec. Publ., 42: 107-124.

Blundy, J.D. and Holland, T.J.B., 1990. Calcic amphibole equilibria and a new amphibole-plagioclase geothermometer. Contrib. Mineral. Petrol., 104: 208-224.

Bonatti, E., Honnorez, J., Kirst, P. and Radicati, F., 1975. Metagabbros from the Mid-Atlantic Ridge at 6°N: Contact-hydrothermal-dynamic metamorphism beneath the axial valley. Journ. Geol., 83: 61-78.

Bonatti, E. and Michael, P.J., 1989. Mantle peridotites from continental rifts to ocean basins to subduction zones. Earth Planet. Sci. Lett., 91: 297-311.

Bonatti, E., Peyve, A., Kepezhinskas, P., Kurentsova, N., Seyler, M., Skolotnev, S., and Udintsev, G., 1992. Upper mantle heterogeneity below the Mid-Atlantic Ridge, 0°-15°N. J. Geophys. Res., 97: 4461-4476.

Boudier, F., 1979. Microstructural studies of three peridotite samples drilled at the western margin of the mid-Atlantic Ridge. Init. Repts. of the DSDP 45: 603-608.

Bougault, H., Cande, S.C., et al., 1985. Initial Reports of the Deep Sea Drilling Project, 82, Washington.

Bougault, H., Charlou, J. L., Fouquet, Y. and Needham, H.D., 1990. Campagne RIDELENTE, structure de la dorsale Atlantique, hétérogenité du manteau et hydrothermalisme. Oceanol. Acta, Vol. Spec., 10: 366-381.

Brown, G.M., 1957. Pyroxenes from the early and middle stages of fractionation of the Skaergaard intrusion, East Greenland. Mineral. Mag., 31: 511-543.

Brown, G.M. and Vincent, E.A., 1963. Pyroxenes from the late stages of fractionation of the Skaergaard intrusion. J. Petrol., 4: 175-197.

Cannat, M., 1993. Emplacement of mantle rocks in the seafloor at mid-ocean ridges. J. Geophys. Res., 98: 4163-4172.

Cannat, M., Bideau, D. and Bougault, H., 1992. Serpentinized peridotites and gabbros in the Mid-Atlantic Ridge axial valley at 15°37'N and 16°52'N. Earth Planet. Sci. Lett., 109: 87-106.

Cannat, M., Juteau, T. and Berger, E., 1991. Petrostructural analyses of the leg 109 serpentinized peridotites. In: R.S. Detrick, J. Honnorez, W.B. Bryan, T. Juteau, et al. (Editors), Proc. of the ODP, Scientific Results, 109: 47-56.

Casey, J.F., 1986. Ultramafic rocks from the MAR at 23° N: evidence for high temperature hydration and high temperature-low to moderate stress deformation of mantle tectonites beneath the median valley. Eos, 67: 1214.

Ceuleneer, G., Nicolas, A. and Boudier, F., 1988. Mantle flow patterns at an oceanic spreading centre: the Oman peridotites record. Tectonophysics, 151: 1-26.

Cormier, M.H., Detrick, R.S. and Purdy, G.M., 1984. Anomalously thin crust in oceanic fracture zones; new seismic constraints from the Kane fracture zone. J. Geophys. Res., 89: 10244-10266.

Detrick, R.S., Fox, P. J., Kastens, K., Ryan, W.B.F., Mayer, L., Karson, J. and Pockalny, R., 1984. A SeaBeam survey of the Kane fracture zone and adjacent Mid-Atlantic ridge rift valley. Eos, 65: 1006.

Detrick, R.S., Honnorez, J., Bryan, W.B., Juteau, T., et al., 1988. Proceeding of the Ocean Drilling Program, Init. Repts., 106/109, College Station, TX.

Dick, H.J.B., 1989. Abyssal peridotites, very slow spreading ridges and ocean ridge magmatism, In: A.D. Saunders and M.J. Norris (Editors), Magmatism in the Ocean Basins. Geol. Soc. London Spec. Publ., 42: 71-105.

Dick, H.J.B., Fisher, R.L. and Bryan, W.B., 1984. Mineralogic variability of the uppermost mantle along mid-ocean ridges. Earth Planet. Sci. Lett., 69: 88-106.

Droop, G.T.R., 1987. A general equation for estimating Fe3+ concentrations in ferromagnesian silicates and oxides from microprobe analyses, using stoichiometric criteria. Mineral. Magazine, 51: 431-435.

Dungan, M.A., 1979. A microprobe study of antigorite and some serpentine pseudomorphs. Canad. Mineral., 17: 771-784.

Engel, C.G. and Fisher, R.L., 1975. Granitic to ultramafic rock complexes of the Indian Ocean ridge system, western Indian Ocean. Geol. Soc. Am. Bull., 86: 1553-1578.

Fisher, R.L., Dick, H.J.B., Natland, J.H. and Meyer, P.S., 1986. Mafic-ultramafic suites of the slowly spreading Southwest Indian Ridge: PROTEA exploration of the Antarctic plate boundary, 24°E-47°E. Ofioliti, 11: 147-178.

Fox, P.J. and Gallo, D.G., 1984. A tectonic model for ridge-transform-ridge plate boundaries: implications for the structure of the oceanic lithosphere. Tectonophysics, 104: 205 - 242.

Francis, T.G., 1981. Serpentinization faults and their role in the tectonics of slow-spreading ridges. J. Geophys. Res., 86: 11616-11622.

Fryer, P., Pearce, J.A., Stokking, L.B., et al., 1990. Proceeding of the Ocean Drilling Program. Init. Repts., 125, College Station, TX.

Gente, P., Zonenshain, L.P., Kuzmin, M., Lisitsin, A.P., Bodganov, Y.A. et Baranov, B.V., 1989. Géologie de l'axe de la dorsale médio-atlantique entre 23° et 26°N: résultats préliminaires de la 15eme campagne du N.O. Akademik Mtislav Keldysh (mars-avril 1988). C.R. Acad. Sci. Paris, 308: 1781-1788.

Gillis, K.M., Thompson, G., and Kelley, D.S., 1993. A view of the lower crustal component of hydrothermal systems at the Mid-Atlantic Ridge. J. Geophys. Res., 98: 19597-19619.

Hamlyn, P.R. and Bonatti, E., 1980. Petrology of mantle-derived ultramafics from the Owen Fracture Zone, Northwest Indian Ocean: implications for the nature of the oceanic upper mantle. Earth Planet. Sci. Lett., 48: 65-79.

Harper, G.D., 1984. The Josephine ophiolite. Geol. Soc. Am. Bull., 95: 1009-1026.

Hébert, R., Serri, G. and Hékinian, R., 1989. Mineral chemistry of ultramafic to gabbroic cumulates from the major oceanic basins and Northern Apennine ophiolites (Italy) - A comparison. Chem. Geol., 77: 183-207.

Juteau, T., Berger, E. and Cannat, M., 1990. Serpentinized, residual mantle peridotites from the M.A.R. Median valley, ODP Hole 670A (21°10'N, 45°02'W, Leg 109): primary mineralogy and geothermometry. In: R.S. Detrick, J. Honnorez, W.B. Bryan, T. Juteau, et al. (Editors), Proc. of the ODP, Scientific Results, 109: 27-45.

Karson, J.A., 1991. Seafloor spreading on the Mid-Atlantic Ridge: implications for the structure of ophiolites and oceanic lithosphere produced in slow-spreading environments. In: J. Malpas, M.J. Moores, A. Panayiotou and C. Xenophontos (Editors), Proc. Symp. Troodos 1987, Geol. Surv. Dep., Nicosia, Cyprus, pp. 547-555.

Karson, J.A. and Dick, H.J.B., 1983. Tectonics of ridge-transform intersections at the Kane fracture zone. Mar. Geophys. Res., 6: 51-98.

Karson, J.A., Collins, J.A. and Casey, J.F., 1984. Geologic and seismic velocity structure of the crust/mantle transition in the Bay of Islands ophiolite complex. J. Geophys. Res., 89: 6126 - 6138.

Karson, J.A., Thompson, G., Humphris, S.E., Edmond, J.M., Bryan, W.B., Brown, J.R., Winters, A.T., Pockalny, R.A., Casey, J.F., Campbell, A.C., Klinkhammer, C., Palmer, M.R., Kinzler, R.J. and Sulanowska, M.M., 1987. Along axis variations in seafloor spreading in the MARK area. Nature, 328: 681-685.

Kirby, S.H., 1983. Rheology of the lithosphere. Rev. Geophys., 21: 1458 - 1483.

Kong, L.S.L., Detrick, R.S., Fox, P.J., Mayer, L.A. and Ryan, W.B.F., 1988. The morphology and tetconics of the Mark area from SeaBeam and SeaMark I observations (Mid-Atlantic Ridge 23° N). Mar. Geophys. Res., 10: 59-90.

Leake, B.E., 1965. The relationship between composition of calciferous amphiboles and grade of metamorphism, In: W.S. Pitcher and G.W. Flinn (Editors), Controls of metamorphism, Edimburgh and London, Oliver & Boyd, pp. 299 - 318.

Leake, B.E., 1978. Nomenclature of amphiboles. Can. Mineral.,16: 501-520.

Liou, J.G., Kuniyoshi, S. and Ito, K., 1974. Experimental studies of the phase relations between greenshist and amphibolite in a basaltic system. Am. J. Sci., 274: 613 - 632.

Manning, C.E. and Bird, D.K., 1986. Hydrothermal clinopyroxenes of the Skaergaard intrusion. Contrib.

Mineral. Petrol., 92: 437-447.
Melson, W.G., Rabinowitz, P.D., et al., 1978. Initial Reports of the Deep Sea Drilling Project, 45, Washington.
Mercier, J.-C. C. and Nicolas, A., 1975. Textures and fabrics of upper-mantle peridotites as illustrated by xenoliths from basalts. J. Petrol., 16: 454-487.
Mével, C., Auzende, J.M., Cannat, M., Dubois, J., Fouquet, Y, Gente, P., Donval, J.P., Grimaud, D., Karson, J.A., Segonzac, M. and Stievenard, M., 1988. HYDROSNAKE 1988: submersible study of seafloor spreading in the MARK area, abstract) AGU, EOS, 69: 1439.
Mével, C., Cannat, M., Pascal, G., Marion, E., Auzende, J.M. and Karson, J.A., 1991. Emplacement of deep crustal and mantle rocks on the west median valley wall of the MARK area (MAR, 23°N). Tectonophysics, 190: 31-53.
Michael, P.J. and Bonatti, E., 1985. Peridotite composition from the North Atlantic: regional and tectonic variations and implications for partial melting. Earth Planet. Sci. Lett., 73: 91-104.
Morris, E. and Detrick, R.S., 1991. Three-dimensional analysis of gravity anomalies in the MARK area (MAR, 23°N). J. Geophys. Res., 96: 4355-4366.
Nicolas, A. and Jackson, M., 1982. High temperature dikes in peridotites: origin by hydraulic fracturing. J. Petrol., 23: 568-582.
Nicolas, A. and Poirier, J.P., 1976. Crystalline plasticity and solid state flow in metamorphic rocks, Wiley, London.
Nicolas, A. and Prinzhofer, A., 1983. Cumulative or residual origin for the transition zone in ophiolites: structural evidence. J. Petrol., 24: 188-206.
Nicolas, A., Reuber, I. and Benn, K., 1988. A new magma chamber model based on structural studies in the Oman ophiolite. Tectonophysics, 151: 87-105.
Nwe, Y.Y., 1976. Electron probe studies of the earlier pyroxenes and olivines from the Skaergaard intrusion, East Greenland. Contrib. Mineral. Petrol., 55: 105-126.
Phillips, J.D., Thompson, G., Bowen, V.T. and Von Herzen, R.P., 1968. Geophysical and geological study of the Mid-Atlantic Ridge. Trans. Amer. Geophys. Union, 49: 327-328.
Pockalny, R.A., Detrick, R.S. and Fox, P.J., 1988. Morphology and tectonics of the Kane transform from seabeam bathymetry data. J. Geophys. Res., 93: 3179-3193.
Prichard, H.M., 1979. A petrographic study of the process of serpentinization in ophiolites and ocean crust. Contrib. Mineral. Petrol., 68: 231-241.
Purdy, G.M. and Detrick, R.S., 1986. Crustal structure of the Mid-Atlantic Ridge at 23°N from seismic refraction studies. J. Geophys. Res., 91: 3739-3762.
Robinson, P., Spear, F.S., Schumaker, J.C., Laird, K., Klein, C., Evans, B.W. and Doolan, B.W., 1981. Phase relations of metamorphic amphiboles: natural occurrence and theory. In: D. Veblen (Editor), Amphiboles and other hydrous pyroboles-Mineralogy, MSA Rev. Mineral., 9B.
Rock, N.M.S. and Carroll, G.W., 1990. MINTAB: a general-purpose mineral recalculation and tabulation program for Macintosh microcomputers. Am. Mineral., 75: 424-430.
Rona, P.A., Widenfalk, L. and Boström, K., 1987. Serpentinized ultramafics and hydrothermal activity at the Mid-Atlantic Ridge crest near 15° N. J. Geophys. Res., 92: 1417-1427.
Schulz, N.J., Detrick, R.S. and Miller, S.P., 1988. Two and three dimensional inversions of magnetic anomalies in the MARK area (Mid-Atlantic Ridge 23°N). Mar. Geophys. Res., 10: 41-57.
Severinghaus, J.P. and Macdonald, K.C., 1988. High inside corners at ridge-transform intersections. Mar. Geophys. Res., 9: 353 - 367.
Sinton, J.M., 1979. Petrology of (alpine-type) peridotites from site 395, DSDP Leg 45. In: W.G. Melson, P.D. Rabinowitz et al. (Editors), Init. Repts. of the DSDP, 45: 595-601.
Sleep, N.H., 1975. Formation of ocean crust: some thermal constraints. J. Geophys. Res., 80: 4037-4042.
Spear, F.S., 1981. An experimental study of hornblende stability and compositional variability in amphibolite. Am., J., Sci., 281: 697-734.
Stakes, D., Mével, C., Cannat, M., and Chaput, T., 1991. Metamorphic stratigraphy of hole 735B. Proc. of the ODP, Scientific Results, 118: 153-180.
Tiezzi, L.J. and Scott, R.B., 1980. Crystal fractionation in a cumulate gabbro, Mid-Atlantic Ridge, 26° N. J. Geophys. Res., 85: 5438 - 5454.
Tucholke, B.E. and Schouten, H., 1988. Kane fracture zone. Mar. Geophys. Res., 10: 1-39.
Vieten, K. and Hamm, H.M., 1978. Additional notes "On the calculation of the crystal chemical formula of clinopyroxenes and their content of Fe3+ from microprobe analyses". Neues Jahrb. Mineral. Monatshefte, 2: 71-83.
Wicks, F.J. and Whittaker, E.J.W., 1977. Serpentine texture and serpentinization. Can. Mineral., 13: 244-258.
Zonenshain, L.P., Kusmin, M.I., Lisitsin, A.P., Bogdanov, Yu.A. and Baranov, B.V., 1989. Tectonics of the Mid-Atlantic rift valley between the TAG and MARK areas (26-24° N): evidence for vertical tectonism. Tectonophysics, 159: 1-23.

Mafic and Ultramafic Intrusions into Upper Mantle Peridotites from Fast Spreading Centers of the Easter Microplate (South East Pacific)

M. CONSTANTIN[1,2], R. HÉKINIAN[1], D. ACKERMAND[3] AND P. STOFFERS[3]

1 "GDR-Genèse et évolution des Domaines océaniques", DRO/Géosciences Marines, IFREMER, Centre de Brest, 29280 Plouzané, France.

2 "GDR-Genèse et évolution des Domaines océaniques", Département des Sciences de la Terre, Université de Bretagne Occidentale, 29285 Brest, France.

3 Geologische-Palaontologisches Institut, Universität Kiel, Olshausenstrasse 40, 2300 Kiel, Germany.

Abstract

Recently, the deepest parts (> 4000-6000 m) of the Easter Microplate bounded by fast spreading ridge segments (11-15 cm/yr total rate) were investigated by the R.V. Sonne. One of these areas is located on the western boundary of the microplate along the West Rift zone near 24°13'S and 115°41'W, and is associated with a small transform fault called Terevaka. The other area consists of a depression called the Pito Deep located at the northern tip of a ridge propagator on the East Rift (near 23°60'S and 111°57'W) and forming the eastern boundary of the microplate. A large variety of crustal rocks were recovered from both localities. These include basalts, diabases, ferrogabbros, gabbros and olivine gabbros. Moreover, an entire suite of mantle harzburgite often showing gabbroic and ferrogabbroic veins and dikelets was found in the Tereveka transform fault. Other ultramafic rocks from Terevaka include wall rock dunitic rims and massive plagioclase dunite resulting from the percolation and impregnation of basaltic liquid in peridotite, and also a rare type of clinopyroxenite veins and clusters made of granoblastic chromium diopside.

Petrological study of samples from the Pito Deep indicates the presence of primitive leucotroctolites (An85, Fo86, clinopyroxene Mg#=89). Harzburgites from the Tereveka transform fault show refractory mineral compositions of spinel (Mg#=61-68, Cr#=40-46), olivine (Fo=91, NiO=0.38%), orthopyroxene (Mg#=91, Al_2O_3=2.5%, Cr_2O_3=0.72%), and trace amounts of clinopyroxene (Mg#=93, Al_2O_3=2.8%, Cr_2O_3=1%, TiO_2<0.05%). Based on their Na content, two varieties of diopside are found in the harzburgite: a low Na-type (Na_2O=0.06%) and a normal Na-type (Na_2O=0.38%). Peridotites impregnated by gabbroic veins show strong chemical disequilibrium with wall rock spinel and clinopyroxene

R.L.M. Vissers and A. Nicolas (Eds.),
Mantle and Lower Crust Exposed in Oceanic Ridges and in Ophiolites, 71-120.

enriched in Fe, Ti and Al contents (spinel: TiO_2=0-6.1%, Cr#=35-68, Mg#=20-65; clinopyroxene: TiO_2=1.1%, Al_2O_3=3.1%, Cr_2O_3=0.8%, Mg#=87).

We present textural and petrological observations for widespread magmatic intrusion, impregnation and melt-solid reaction between transient basaltic melts and refractory mantle peridotites formed at a fast spreading center. This study describes in detail magmatic features which we believe are part of the accretion process occurring within the lower crust-upper mantle transition zone. Comparisons with other Pacific regions suggest that similar magmatic processes of melt impregnation and reaction occur in a variety of tectonic settings such as transform faults (i.e., Tereveka, Garrett) and propagating rifts (i.e., Pito Deep, Hess Deep). It also reinforces the evidence that interaction of basaltic melt with peridotite residues is a prominent phenomenon in the genesis of oceanic lithosphere.

Introduction

Limited information exists about upper mantle and lower crustal rocks generated at fast and very fast spreading centers. We present the first detailed petrologic study describing upper mantle and lower crustal rocks

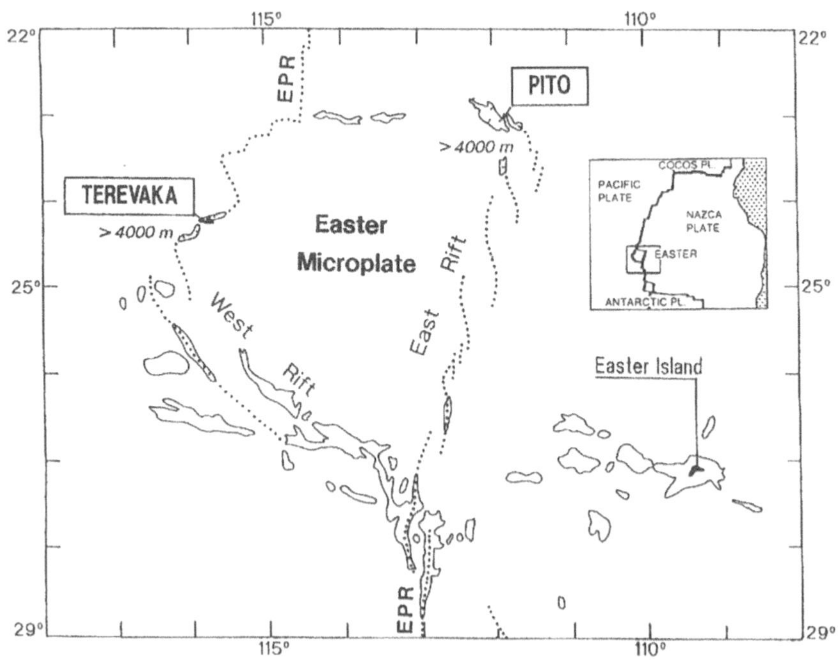

Figure 1. Regional map of the Easter microplate, showing locations of the Terevaka transform fault and Pito Deep. Dotted lines represent actively accreting ridge segments (after Naar and Hey, 1991).

formed along the Easter microplate in the Eastern Pacific Ocean (Fig. 1). This work describes the various mafic and ultramafic rocks which are found on the floor and walls of two very deep (> 5000 m) sites, the Pito Deep and the Terevaka transform fault.

The Terevaka transform fault is one of the few localities where widespread magmatic intrusion, percolation and reaction of basaltic liquid into abyssal peridotite have been recognized (Constantin et al., 1993). Gabbroic and ultramafic rocks have previously been described from a few areas of the South-East Pacific region. The two main locations along the East Pacific Rise (EPR) are the Garrett transform fault at 13°30'S (Hébert et al., 1983, 1989; Cannat et al., 1990; Hékinian et al., 1992) and the Hess Deep at 2°15'N (Francheteau et al., 1990; Hékinian et al., 1993; Girardeau and Francheteau, 1993). A more complete and thorough comparison of the regional mineral characteristics of these three areas is the subject of current study. Gabbroic dikelets and veins in oceanic mantle peridotites are scarce and usually of local extent. They have been reported from two Mid-Atlantic Ridge (MAR) rift valley localities at 15°37'N and near the Kane fracture zone at 23°N (Mével et al., 1991; Cannat et al., 1992), and from various fracture zones of the South-West Indian Ridge (Engel and Fisher, 1975; Fisher et al., 1986; Bloomer et al., 1989). Known slow-spreading ridge segments or associated fracture zones where plagioclase-bearing peridotites have been found include some fracture zones of the South-West Indian Ridge (SWIR) and the America-Antarctica ridge, and the Romanche fracture zone (Dick, 1989; Bonatti et al., 1992). Although clinopyroxenites are frequent in mantle sections of ophiolite complexes, their recovery in ocean basins is extremely rare. In addition to this study, reported occurrences of clinopyroxenites with unrelated continental affinities come from the Indian ocean in the Ob trench (Hékinian, 1970) and southwest of Australia (Nicholls et al., 1981), and also from the Norwegian-Greenland sea (Bailey et al., 1992).

This paper deals primarily with the petrology of gabbros and peridotites recovered during the R.V. Sonne cruises 65 and 80 (Stoffers et al., 1989; 1992) in the Easter microplate, southeast Pacific. Several dredge hauls and deep-towed camera runs were performed in the Pito Deep located at the tip of a propagator near 22°59'S and 111°56'W, and in the Terevaka transform fault at 24°13'S and 115°41'W. All the oceanic rock types mentioned above are found at a single locality in the Terevaka transform fault

Geological setting

Most previous studies were concerned with structural and geophysical surveys focussed on the geometry and regional tectonic fabric of the Easter microplate, devoted to better comprehend its evolution with respect to the EPR (Hey et al., 1985; Naar and Hey, 1991; Rusby, 1992; Searle et al.,

Table 1. Petrography of dolerite and massive gabbroic rocks from Pito Deep and Terevaka transform fault in the Easter microplate.

Rock types	*Pito Deep*		*Terevaka transform fault*	
	Sample (SO65-)[1]	*Primary mineralogy*[2]	*Sample (SO80-83DS)*[1]	*Primary mineralogy*[2]
Diabase	09DS-01, 11DS-03	pl, cpx, il An_{60}, $Mg\#_{76}$, Wo_{35}	34, 38, M13	pl, cpx $An_{49,54,63,74}$, $Mg\#_{76}$, $Wo_{34,42}$
Ferrogabbro	X	X	18, 35, 40, M17	pl, cpx, opx, il $An_{33,45,47,51,61}$, $Mg\#_{63,68,81}$, $Wo_{41,42}$, $Mg\#_{54}$, Wo_3
Metagabbro	X	X	7, 30, M13	pl, cpx An_{81}, $Mg\#_{76}$, Wo_{45}
Gabbro	18DS-03, 18DS-05, 18DS-09, 18DS-10, 18DS-11	pl, cpx, trace ol, trace opx An_{65-66}, $Mg\#_{78-80}$, Wo_{41-45}, Fo_{73}, $Mg\#_{76}$, Wo_2	32, M13	pl, cpx $An_{66,81}$, $Mg\#_{84}$, Wo_{43}
Microgabbro	18DS-06, 18DS-10, 18DS-11	47%pl, 49%cpx, 3% opx, 1%ol An_{66-69}, $Mg\#_{80-81}$, Wo_{42-44}, Fo_{75}, $Mg\#_{76-77}$, Wo_{2-3}	X	X
Olivine gabbro	12GTV, 18DS-04, 18DS-07, 18DS-08, 18DS-12	56%pl, 37%cpx, 7%ol, trace opx An_{65-69}, $Mg\#_{80-83}$, Wo_{41-44}, Fo_{73-78}, $Mg\#_{77-78}$, Wo_2	X	X
Leucotroctolite	11DS, 11DS-01, 11DS-02	81%pl, 8%ol, 1%cpx, trace opx An_{85-86}, Fo_{86-88}, $Mg\#_{89-91}$, Wo_{45-46}, $Mg\#_{86-88}$, Wo_1	X	X

[1] R.V. Sonne 65 and 80 cruises. Subset (-34, -38...) indicates sample number, DS the dredge number. Samples in boldface are representative samples for which modal point counting is presented in adjacent column.

[2] Mineral and chemical data listed in decreasing order of modal abundance. Modal point counting performed at 1mm spacing for 1000-2000 point counts.

1993). Sea Beam bathymetry, SeaMARC II side-scan sonar, gravity and magnetic investigations of the Pito area were reported by Francheteau et al. (1988), Naar et al. (1991), and Martinez et al. (1991). The Easter microplate has an average diameter of 400 km, and is bounded along its eastern and western margin by two sets of accreting ridge segments: the East rift and the West rift (Fig. 1). The northern (23°S) and southern (27°S) boundaries of the microplate are made up of shallow ridges (i.e., the Pito and Orongo ridges) representing faulted structures (Francheteau et al., 1988; Searle et al., 1989). These large transform faults have offset this section of the very fast spreading EPR (120-160 km/Ma total rate, according to DeMets et al., 1990). Naar and Hey (1991) followed by Searle et al. (1993) demonstrated that the microplate results from the complex boundary reorganization of the two large and neighbouring Nazca and Pacific plates. They also showed that all of the microplate boundaries are volcanically and tectonically active and probably change configuration as a consequence of the microplate's rapid (15°/Ma) clockwise rotation. The two accreting ridge systems bounding the microplate differ from each other in the style of their tectonic settings. The northern part of the East rift is a northward propagator which is faulting and stretching the pre-existing (3 Ma) Nazca lithosphere (Martinez et al., 1991). This propagator terminates near 23°00'S and 111°57'W in a depression called the Pito Deep (5985 m depth). The area is extensively faulted by northwest-southeast oriented scarps, frequently 1000 m high and forming 4000 m of relief (Naar et al., 1991). Whilst the entire length of the East Rift consists essentially of rift propagator and overlapping spreading centers, the West rift is offset by several small transform faults (Naar and Hey, 1991). One of these structures, the Terevaka transform fault (Rusby, 1992) located from 24°06'S-115°26'W to 24°30'S-116°10'W, was the site of a bathymetric survey and dredging attempts (Stoffers et al., 1992). This left lateral transform fault is made of two sub-parallel valleys, 40 and 55 km long (total offset of 100 km) and almost 5000 m deep, and has a typical morphology of a medium to fast-slipping, long-offset transform fault (Rusby, 1992, after Fox and Gallo, 1984). Naar and Hey (1989) calculated a full spreading rate of 118 km/Ma for the ridge segment located north of the Terevaka transform fault. Using the perpendicular distance between the maximum age contrast represented by the 1.77 Ma magnetic anomaly north of the West Rift-transform fault intersection and the present-day active ridge, we estimate a slip rate of approximately 113 km/Ma.

Geochemical characteristics of dredged volcanic rocks along various active segments of the microplate were published by Schilling et al. (1985), Hanan and Schilling (1989), Fontignie and Schilling (1991), and Poreda et al. (1993). They showed with trace element content and isotopic composition that basalts dredged along the West Rift have normal MORB characteristics. On the other hand, the middle part (near 26°S-111°W) of the East Rift consists of enriched MORB which the authors assign to the influence

Table 2. Petrography of intrusive rocks in harzburgite from the Terevaka transform fault, Easter microplate.

Type of intrusion[1]	*Sample*[2] *(SO80-83DS-)*	*Magmatic mineral chemistry*[3]	*Vein texture and width*	*Rock description*
Clinopyroxenite	M28a, M33, M163	cpx, opx, ol, sp Mg#$_{92}$, Wo$_{46}$, Fo$_{89,91}$, Mg#$_{62}$, Cr#$_{47}$, Mg#$_{87,91}$, Wo$_{1,5}$	Granoblastic 2-5mm	Band set parallel to host harzburgite foliation. Also fine grained cluster located near gabbro impregnation zone.
Plagioclase-dunite	M11, M18, M153	pl, ol, sp, cpx, opx, kaer An$_{68,82}$, Fo$_{86-88}$, Mg#$_{46,56}$, Cr#$_{52-55}$, Mg#$_{90-92}$, Wo$_{43-45}$, Mg#$_{89}$, Wo$_{3}$, Mg#$_{70-73,88}$	Granular 2-3mm	Host dunite with heterogeneous distribution (<15%) of interstitial, elongate and coalescent plagioclase grains.
Dunite veinlet	26, 42	ol Fo$_{90-91}$, NiO=0.24%	Granoblastic 1mm	Dunitic veinlet located within host dunitic harzburgite, discordant with host foliation and with gabbroic contact.
Olivine microgabbro	42, M09	pl, cpx, ol, kaer An$_{58,65}$, Mg#$_{82-84}$, Wo$_{44-45}$, Fo$_{75}$, Mg#$_{85}$	Equigranular >=15mm	Dikelets and veins are set parallel to wall rock and host harzburgite foliation. Very fine grained: injection-crys tallization followed by mineral flow.
Gabbro	08, 24, 26, M09, M15	pl, cpx An$_{51,55,66}$, Mg#$_{83-91}$, Wo$_{38,43-45}$	Granular 7-50mm	Injection veins: minerals are set parallel to wall rock. Undeformed tension veins set perpendicular or at 45° discordant with host harzburgite to dunite foliation.
Mylonitic gabbro	28, M01, M14 M19, M39b	pl, cpx, +-opx An$_{58-60}$, Mg#$_{88-93}$, Wo$_{44-48}$	Mylonitic to gneissic 2-10mm	Mylonitic veins are parallel to host harzburgite foliation Intrusion concurrent with synkinematic deformation.
Ferrogabbro	24, M12, M21, M31, M36	pl, cpx, kaer, il, ap, zir, An$_{26,46}$, Mg#$_{40,46}$, Wo$_{38,40}$ Mg#$_{87}$	Cataclastic to porphyroclastic 5-20mm	Injection veins and pockets. Medium-coarse grained. Foliation parallel to wall rock but discordant with host harzburgite foliation.
Mylonitic ferrogabbro	M16, M25, M28, M153	pl, cpx, kaer, il, ap, zir An$_{33,45-48,55}$, Mg#$_{49,65-75,81}$, Wo$_{39-44}$	Mylonitic to gneissic 2-10mm	Multiple self-injections, highly deformed, fine grained. Foliation and banding are parallel to wall rock but dis cordant with host dunite foliation.

1 Intrusive rock types listed in inferred order of magmatic appearance. All rock types form intrusive veins, lenses or pockets in peridotite.
2 R.V. Sonne 80 cruise, dredge site 83. The subset (M28, 26..) indicates the sample number in the dredge haul.
3 Mineral and chemical data are listed in decreasing order of modal abundance. Only analyses performed in middle of veins or dikelet were used.

of the nearby Easter Island hot spot. To the best of our present knowledge, the two areas investigated in this study are located outside the influence of such a hot spot.

Petrography

Pito Deep

Four sampling runs (dredges and television-grab) were conducted between 5800 and 5250 m water depth at 22°59.65'-22°59.85'S and 111°55.81'-111°56.66'W (Table 1). Rock types recovered include diabase, ferrogabbro, amphibolitized gabbro, microgabbro, olivine gabbro, and leucotroctolite. The most abundant rock type recovered is a very fresh, fine- to medium-grained, foliated granular olivine gabbro. Although each sample is homogeneous, small modal and grain size variations exist between the various gabbroic types. Leucotroctolites, found in dredge 11 from the Pito Deep, contain euhedral medium-grained plagioclase and olivine, with small interstitial or poikilitic clinopyroxene. The leucotroctolites have a homogeneous and a more primitive composition than the olivine gabbro. The compositional homogeneity suggests that the mineralogical variations observed in these massive gabbroic rocks are the result of varying conditions of mineral nucleation and fractional crystallization. It is also believed that viscous flow of the olivine gabbro crystalline mush produced the foliation and led to textural homogenisation during lithospheric spreading (Benn and Allard, 1989; Nicolas, 1992). The sampled section may correspond to the dike complex and layer 3 of the oceanic lithosphere.

Terevaka transform fault

One dredge station (83DS) was made in the Terevaka transform fault (24°12.91'S-24°14.32'S and 115°40.72'W-115°41.40'W), going from the transform valley floor and up along the southern wall between 4850-4200 m (Fig. 1). A wide range of relatively fresh rocks with remarkably well preserved magmatic structures and textures were collected. The volume proportions of rocks, based on more than 250 samples from this dredge, are 8% diabase, 15% massive gabbroic rocks with equal proportions of ilmenite-bearing gabbros (ferrogabbros) and strongly amphibolitized gabbros (metagabbros), and 77% peridotites (predominantly harzburgite and subordinate dunite) including 35% with gabbroic veins, 22% vein-free peridotites and 20% strongly serpentinized and altered peridotites.

The diabase samples have fine-grained doleritic textures. Massive ferrogabbros have heterogeneous brecciated textures, with coarse-grained (up to 5 cm) clinopyroxene and medium-grained, subhedral plagioclase set in an amphibole- and ilmenite-rich interstitial matrix. One sample was found of a foliated massive gabbro (83DS-32) with an equigranular texture,

Plate 1. Intrusion and impregnation in ultramafic rocks from the Terevaka transform fault (SO80-83DS dredge). Arrows point to lithological contacts.

(A) Clinopyroxenite vein oriented parallel to harzburgite foliation (sample 83DS-M163). Note sharp but wavy contacts and the dunitic character of the wallrock margins (black halos). Both lithologies are cut at high angle by a metamorphosed gabbro vein at the right hand side (arrow); (B) Strongly serpentinized peridotite (sample 83DS-M12) with anastomosing serpentine and magnetite veins (to the right), in contact with porphyroclastic apatite-bearing, olivine-free ferrogabbro pocket (to the left). Two types of reaction zones are developed between the two lithologies: first a 1.7 cm-thick black dunitic layer and second a 0.6 cm-thick olivine- and ilmenite-rich ferrogabbro layer (detail shown in Plate 3C); (C) Plagioclase dunite (sample 83DS-M11) showing regular patchy distribution of interstitial, oriented plagioclase grains (white). This texture is thought to result from basalt impregnation in peridotite. Note late plagioclase-rich veinlets radially injected from lower right of sample. Detail of microstructure shown in Plate 3A; (D) Serpentinized harzburgite (sample 83DS-42) in sharp and parallel contact with olivine microgabbro dikelet (1.6 cm-thick). Late high-angle tension fractures are filled with white carbonate; (E) Plagioclase-rich percolation vein (white) in dunite (sample 83DS-M18) illustrating how plagioclase infiltrates the peridotite. Note discontinuous nature of the vein and its high angle to the host rock foliation. White tiny veinlets are filled with serpentine; (F) Sample (83DS-26) showing two wavy contacts (centre and upper edge) between peridotite (black) and porphyroclastic gabbro (light grey). Peridotite contains trails, parallel to the contacts with the gabbro, of intensely stretched lenticular orthopyroxenes with recrystallized olivine. A detail of the peridotite microstructure is shown in Plate 3B; (G) Porphyroclastic to mylonitic harzburgite (sample 83DS-28) with mylonitic gabbro vein (in upper part). Host rock foliation, vein minerals and vein wall rock are all subparallel due to high- temperature plastic deformation. Plagioclase exclusively occurs inside the vein (top part). Late carbonate-filled tension cracks are oriented oblique with respect to host rock foliation.

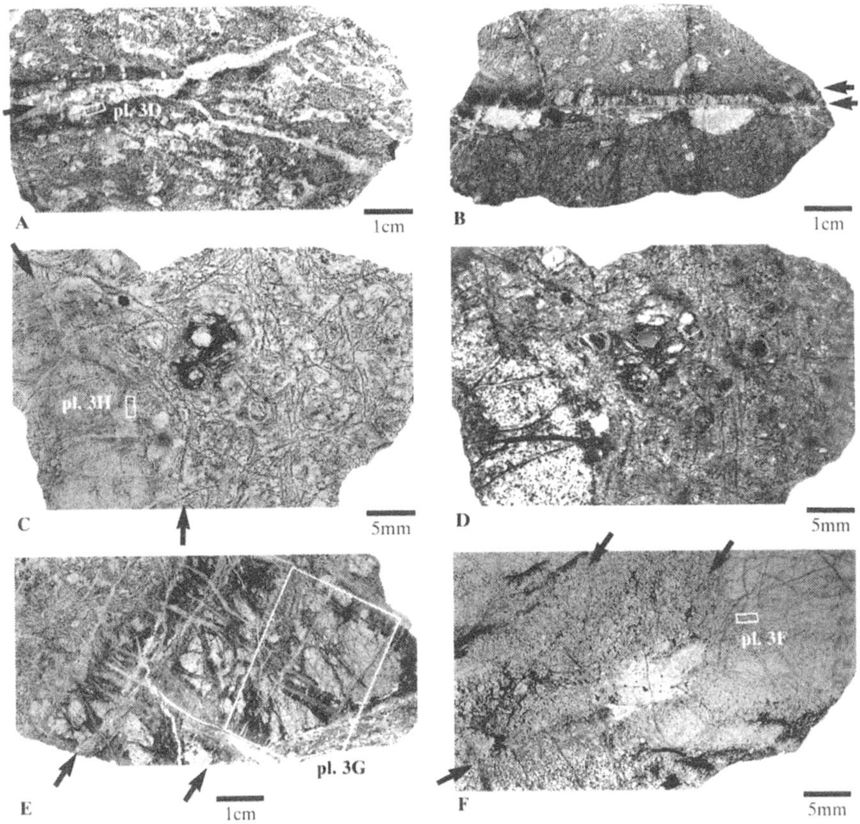

Plate 2. Whole-section photomicrographs of intrusion and impregnation in ultramafic rocks from the Terevaka transform fault (SO80-83DS dredge). Arrows point to lithological contacts.

(A) Tip of intrusive apatite- and zircon-bearing metagabbro vein in harzburgite (sample 83DS-M25). Plagioclase and clinopyroxene of gabbro vein are replaced by amphibole, chlorite and talc. Note thet the orientation of the vein relative to the foliation in the surrounding rock is consistent with a hydraulic fracturing-injection mechanism. Host peridotite contains dunitic halos along contact with the widest part of the vein (left). Detail as outlined shown in Plate 3D; (B) Ferrogabbro intrusion (lower part) showing dunitic reaction front along harzburgite (upper part) wall rock (sample 83DS-M21). Minerals on ferrogabbro side of contact are completely sheared by the flow. The ferrogabbro contains a partly corroded, oval-shaped dunitic clast, 6 mm long and surrounded by opaques, which appears detached from the reaction front and entrained within the clinopyroxene-rich ferrogabbro. Dunitic front and host harzburgite are cut at high angles by two veinlets rooting in plagioclase- and brown amphibole-rich band along sheared contact zone; (C) Clinopyroxenitic cluster showing sharp but wavy contact with host harzburgite (sample 83DS-M28). Left part is made up of equigranular fine-grained diopside, middle and right parts are porphyroclastic host harzburgite. Plane polarized light; detail as outlined shown in Plate 3H; (D) same as (C), nicols crossed; (E) Peridotite intruded by multiple sets (arrows) of ferrogabbroic injection veins (sample 83DS-M29). Note typical host harzburgite free of veins and patches in upper left corner of sample. Branched geometry of ferrogabbroic veins suggests injection from bottom side towards upper right. Magmatic veins are cut at high angles by late serpentine-filled tension veins (running from upper left to lower right). Domain in box shown in Plate 3G contains patches of recrystallized peridotitic wall rock; (F) Clinopyroxene-rich rock (sample 83DS-M33) showing gradual modal and compositional variation suggesting a sequence of metasomatic events. Plane polarized light; same view shown with nicols crossed in Plate 3E. A 3 cm long traverse starting symmetrically from the borders (upper left and lower right) comprises medium- to coarse-grained, plastically deformed Cr- and Al-rich diopside coexisting with holly-leaf spinel and stretched and partly disaggregated olivine clusters. Diopside in turn is partly corroded and replaced by abundant small foliated tablet-shape augite granoblasts. Mineral assemblage in the central part of sample (delimited by arrows) is composed of augite + plagioclase + kaersutite + interstitial sulphides. For further explanation see text. Detail as outlined shown in Plate 3F.

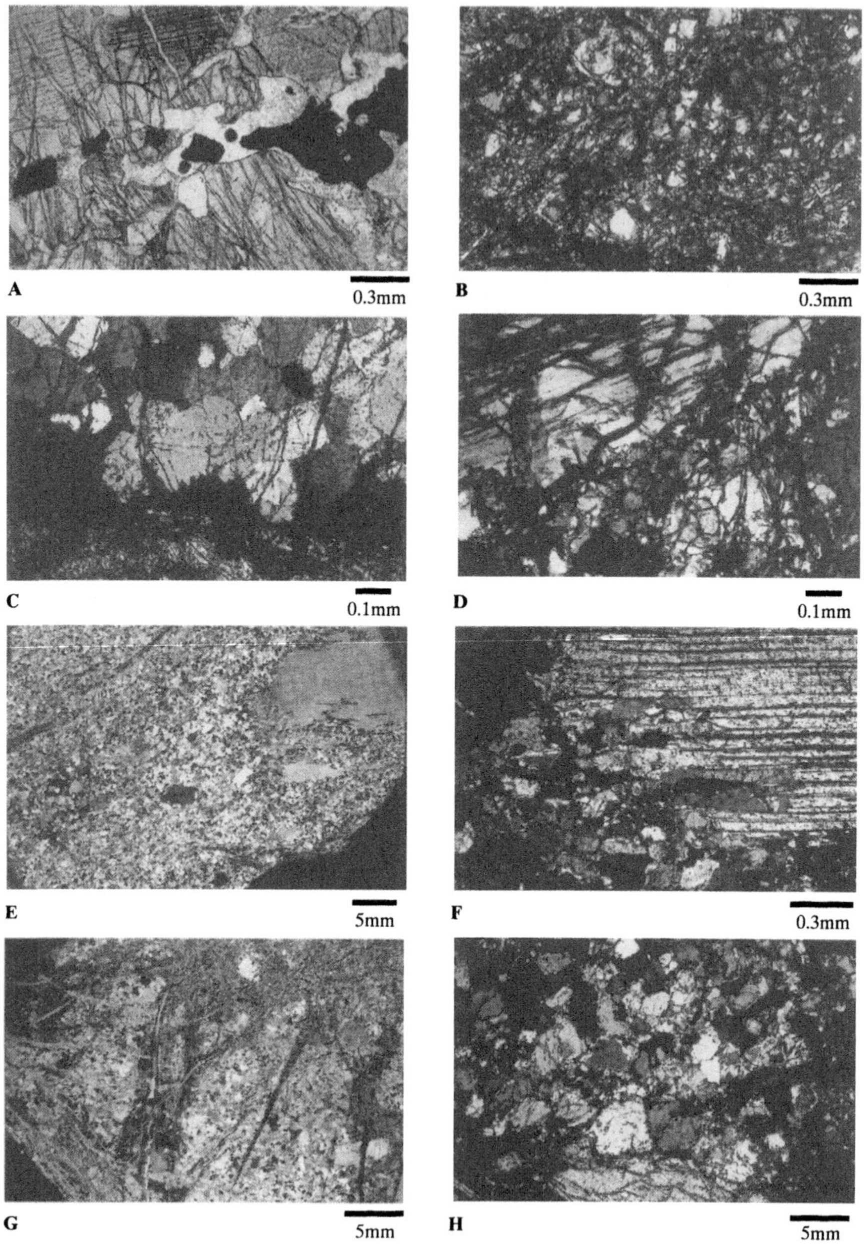

Plate 3. Photomicrographs of intrusive and reactional textures in ultramafic rocks from the Terevaka transform fault (SO80-83DS dredge).

(A) Elongate interstitial plagioclase grains and associated clinopyroxene in dunitic matrix (sample 83DS-M11). Note that plagioclase grains include trails of spinel; (B) Dunitic veinlet cutting through recrystallized dunitic harzburgite (sample 83DS-26). The veinlet is 0.2 mm-thick, runs from upper right to lower left and shows simultaneous extinction. This veinlet is subparallel to the lower gabbro contact of Plate 1F; (C) Reaction layer (0.6 cm thick) in ferrogabbro/peridotite contact zone of Plate 1B. Upper part made up of comparable to those of Pito Deep. Aside the numerous dikelets and veins in the peridotites, few massive olivine gabbro were found. The peridotites recovered are partially to completely (40 to 100%) metamorphosed due to pervasive serpentinization of primary minerals (70.5% olivine + 28% orthopyroxene + 1.5% spinel + trace clinopyroxene by volume in sample 83DS-20). Between 40 and 60% of the peridotites recovered in the dredge haul (83DS) show veins and pods of gabbros and traces of pyroxenite. Where in contact with gabbro, most of the peridotites have generally dunitic rims or patches showing specifically fresh olivine-rich aureoles. On the basis of structural, textural and mineralogical criteria, several types of gabbroic veins and dikelets intrusive into the peridotite have been distinguished (Table 2). The majority of the veins are intrusive gabbros, which either are rotated by plastic deformation and set parallel to the host peridotite foliation, or are undeformed and crosscut the foliation trend (Plates 1G and 2A, respectively). In the deformed veins, primary minerals (plagioclase + clinopyroxene) show major shearing, crushing and abundant recrystallization. In the undeformed veins, minerals inside the veins are set parallel to the wall rock contact. The intrusive gabbroic rocks include olivine-bearing microgabbro dikelets (Plate 1D), gabbro (Plate 1F), ferrogabbro and metagabbro intrusive pods (Plates 1B, 2A, 2B), with sometimes well-preserved patchy peridotitic rims (Plates 2B, 2E). A rare lithological type consists of clinopyroxenite, forming veins or patches made of granoblastic chromium-rich diopside (Plates 1A, 2C, 2F). Finally, foliated and mottled plagioclase dunites were also identified (Plates 1C, 1E).

The description of the various structures observed in the intruded peridotites (Table 2) involves problems in the choice of appropriate petrographic terms free of genetic connotation. For example, the term intrusion designating the igneous rock mass formed within the surrounding peridotite, refers more specifically here to crosscutting magmatic veins and dikelets. These latter are mostly discordant to the foliation, except for cases where plastic deformation has rotated the veins into parallelism with the foliation. These intrusions vary in size, from a few tens of mm up to 5 cm in thickness, and show sharp contacts marking confined interaction with the host peridotite (Plates 1D, 1G, 2A). Instead, in the case of lentic-

---

polygonal olivine (Fo27) embayed in Fe-Ti oxides (central part), lower part shows partly altered plagioclase (An27) from ferrogabbro (83DS-M12); (D) Detail of metagabbro-harzburgite contact zone (sample 83DS-M25) showing deformed enstatite crystal, partly replaced by small recrystallized, polygonal olivine crystals; (E) Same as Plate 2F, nicols crossed; (F) Detail of Plate 2F showing deformed Cr- and Al-rich diopside crystal, corroded along its cleavage by abundant augite granoblasts (sample 83DS-M33); (G) Detail of Plate 2E showing recrystallized peridotite wall rock adjacent to intrusive ferrogabbro (sample 83DS-M29). Top part shows strongly serpentinized peridotite (grey); lower left part includes an amphibole-rich ferrogabbro vein. Note granoblastic recrystallization texture (central part) of the peridotite wall rock, made up of relatively fine-grained, anhedral orthopyroxene and olivine with tiny inclusions of spinel; (H) Detail of Plate 2C of contact zone between clinopyroxenite and host harzburgite (sample 83DS-M28), showing a large enstatite porphyroclast (lower edge) replaced by equigranular, fine-grained diopside and minor amounts of granoblastic enstatite.

Table 3. Microprobe analyses of spinel in ultramafic rocks from the Terevaka transform fault.Rock types include PER: host harzburgite, PLDU: plagioclase dunite, CPXN: clinopyroxenite, G: gabbro, v: vein in peridotite, z: wall rock reaction zone adjacent to contact, c: cryptic zone (≤ 1cm from intrusion), g: granoblast.
avg: average number of microprobe analyses, lithol: associated lithology, distance: in mm from contact
Mg#=100*Mg/(Fe2+Mg), Cr#=100*Cr/(Cr+Al), Fe3#=100*Fe3/(Fe3+Cr+Al)

Sample	*24*	*26*	*28*	*42*	*42*	*43*	*43*	*M11*	*M12*	*M12*	*M12*	*M14*	*M14*	*M14*
Rock	*PERz*	*PERc*	*PER*	*PERc*	*PERz*	*PERg*	*PER*	*PLDU*	*PERc*	*PERz*	*PERzg*	*PERz*	*PERc*	*PERc*
avg	*10*	*8*	*15*	*5*	*3*	*2*	*7*	*17*	*5*	*4*	*5*	*2*	*6*	*3*
lithol	*FGv*	*Mg*	-	*MIGOv*	*MIGOv*	-	-	-	*FGv*	*FGv*	*FGv*	*MYGv*	*MYGv*	*MYGv*
distance	*<1*	*15*	-	*8*	*<1*	-	-	-	*20*	*12*	*15*	*1*	*7*	*10*
TiO_2	5.32	0.06	0.01	0.25	1.11	0.01	0.02	0.96	0.02	0.07	0.03	0.46	0.03	0.09
Al_2O_3	14.34	28.92	31.17	32.06	25.49	35.09	30.48	23.14	33.23	32.35	33.75	29.27	30.35	20.75
Cr_2O_3	31.82	37.92	37.78	35.82	40.16	32.11	38.14	42.47	37.82	38.18	34.92	38.23	40.23	50.53
FeO	39.17	17.31	16.79	16.89	20.10	17.93	16.61	19.48	14.12	14.29	17.57	17.66	15.10	17.64
MnO	0.39	0.18	0.18	0.18	0.19	0.20	0.16	0.22	0.25	0.20	0.22	0.24	0.20	0.30
MgO	6.81	14.31	14.80	13.61	11.43	14.64	14.69	12.18	15.56	15.42	13.73	14.31	15.10	12.47
Total	97.95	98.76	100.80	98.85	98.54	100.01	100.31	98.55	101.07	100.54	100.27	100.19	101.08	101.79
O=32														
Ti	1.074	0.010	0.003	0.044	0.206	0.002	0.003	0.181	0.003	0.012	0.005	0.082	0.005	0.016
Al	4.540	8.169	8.564	8.980	7.455	9.561	8.425	6.795	9.003	8.841	9.282	8.165	8.335	5.983
Cr	6.756	7.185	6.967	6.739	7.881	5.873	7.073	8.373	6.875	6.999	6.447	7.154	7.410	9.777
Fe_3	2.556	0.623	0.466	0.194	0.249	0.561	0.483	0.470	0.112	0.139	0.264	0.522	0.242	0.209
Fe_2	6.247	2.846	2.809	3.170	3.923	2.908	2.776	3.594	2.601	2.631	3.167	2.974	2.700	3.401
Mn	0.089	0.036	0.035	0.036	0.041	0.038	0.032	0.047	0.048	0.040	0.043	0.048	0.039	0.061
Mg	2.722	5.110	5.143	4.822	4.227	5.047	5.137	4.526	5.333	5.328	4.777	5.049	5.244	4.546
Mg#	30.32	64.23	64.67	60.34	51.86	63.45	64.92	55.75	67.22	66.94	60.14	62.93	66.01	57.21
Cr#	59.83	46.79	44.86	42.88	51.39	38.05	45.64	55.20	43.30	44.19	41.00	46.70	47.07	62.04
Fe3#	18.42	3.90	2.91	1.22	1.60	3.50	3.02	3.00	0.70	0.87	1.65	3.29	1.51	1.31

Table 3 continued. Microprobe analyses of spinel in ultramafic rocks from the Terevaka transform fault.

Sample	*M14*	*M18*	*M18*	*M21*	*M21*	*M28*	*M28*	*M28*	*M28*	*M33*	*M163*	*M163*	*M163*	*M163*
Rock	*PERc*	*PERz*	*PLDUv*	*PERc*	*PERz*	*Gvz*	*PERc*	*PERc*	*CPXN*	*CPXN*	*CPXN*	*PERc*	*PERc*	*PERz*
avg	*3*	*9*	*4*	*4*	*1*	*1*	*3*	*9*	*3*	*6*	*6*	*5*	*3*	*11*
lithol	*MYGv*	*PLDUv*	-	*FGv*	*FGv*	-	*FGv*	*CPXN*	-	*Gz*	-	*CPXN*	*CPXN*	*CPXN*
distance	*10*	*10*	-	*20*	*5*	-	*5*	*10*	-	*11*	-	*7*	*7*	*1*
TiO_2	0.09	0.09	0.84	0.07	0.01	0.58	0.08	0.17	0.05	0.28	0.49	0.02	0.02	0.15
Al_2O_3	20.75	26.56	23.82	33.91	34.65	22.28	33.88	37.22	42.63	29.84	24.15	32.94	38.25	34.12
Cr_2O_3	50.53	42.24	44.03	36.70	34.91	44.65	35.03	33.15	26.99	39.17	41.31	36.48	29.59	34.31
FeO	17.64	18.64	19.08	14.62	15.50	19.87	14.02	13.12	13.53	17.37	18.57	14.07	13.99	14.50
MnO	0.30	0.19	0.15	0.17	0.03	0.25	0.28	0.15	0.19	0.22	0.23	0.22	0.22	0.19
MgO	12.47	12.46	12.35	15.74	15.21	11.14	15.59	16.66	16.74	14.04	12.36	15.25	15.96	15.31
Total	101.79	100.30	100.37	101.30	100.41	98.84	98.94	100.61	100.46	100.96	97.57	99.09	98.10	98.71
O=32														
Ti	0.016	0.017	0.154	0.011	0.002	0.110	0.013	0.029	0.008	0.049	0.091	0.003	0.004	0.026
Al	5.983	7.563	6.863	9.134	9.402	6.597	9.306	9.904	11.123	8.268	7.092	9.097	10.378	9.401
Cr	9.777	8.100	8.515	6.632	6.354	8.870	6.455	5.917	4.725	7.285	8.154	6.759	5.384	6.343
Fe_3	0.209	0.300	0.308	0.204	0.237	0.312	0.210	0.119	0.126	0.346	0.560	0.139	0.231	0.200
Fe_2	3.401	3.479	3.594	2.590	2.747	3.863	2.523	2.362	2.379	3.070	3.322	2.619	2.461	2.635
Mn	0.061	0.040	0.031	0.033	0.006	0.054	0.056	0.030	0.036	0.045	0.048	0.043	0.043	0.037
Mg	4.546	4.488	4.500	5.360	5.218	4.171	5.417	5.608	5.525	4.921	4.589	5.328	5.477	5.335
Mg#	57.21	56.33	55.60	67.42	65.51	51.92	68.22	70.37	69.90	61.58	58.00	67.04	68.99	66.94
Cr#	62.04	51.70	55.36	42.06	40.33	57.35	40.96	37.40	29.82	46.84	53.49	42.63	34.16	40.29
Fe3#	1.31	1.88	1.96	1.28	1.48	1.98	1.31	0.74	0.79	2.18	3.54	0.87	1.44	1.25

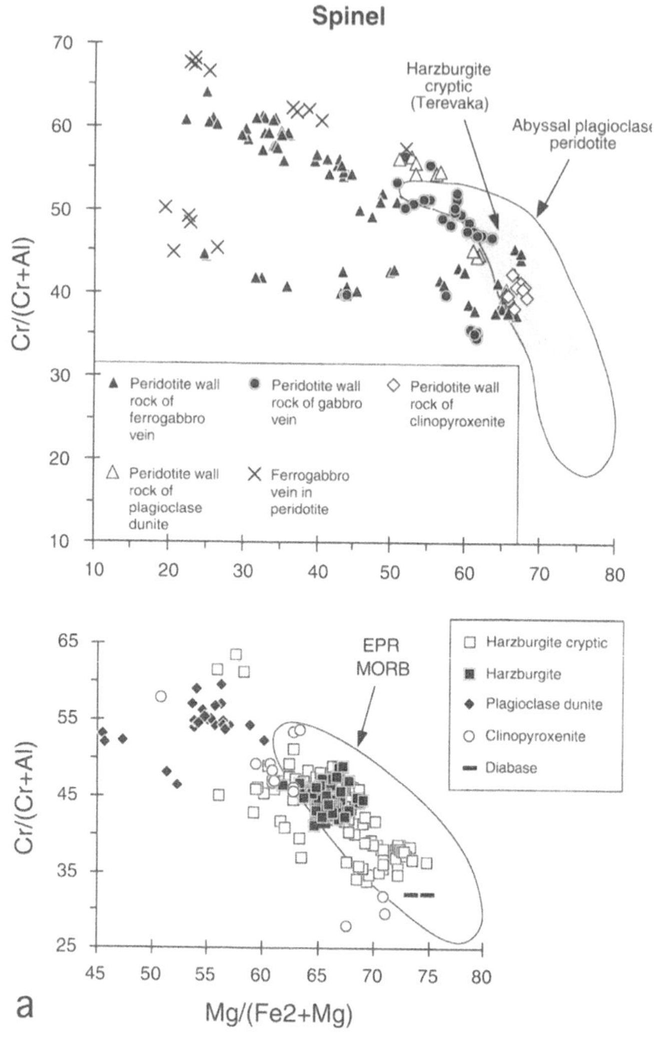

Figure 2. Compositional variations of spinels from the Terevaka transform fault samples: (a) variation diagrams of Cr# versus Mg#. Upper diagram shows spinel in recrystallized peridotitic wall rock adjacent to intrusive rocks. Spinel from peridotite wallrock adjacent to ferrogabbro intrusive veins show two trends which both extend to very low Mg#. Cryptic harzburgite field refers to spinel from unrecrystallized harzburgite wall rock portion located within 1 cm of the intrusive rock and away from the immediate contact. Abyssal plagioclase-peridotite field after Dick and Bullen (1984). Lower diagram shows spinel from ultramafic rocks. Field of East Pacific Rise (EPR) MORB compiled from a total of 57 spinels (unpublished analyses) in samples from the 13°N ridge segment, the Garrett transform fault, and Hess Deep.

ular shaped, patchy zones containing irregularly distributed plagioclase and clinopyroxene (Plate 1C), and in the case of interstitial plagioclase growth (Plate 1E), the general term plagioclase-bearing peridotite is used.

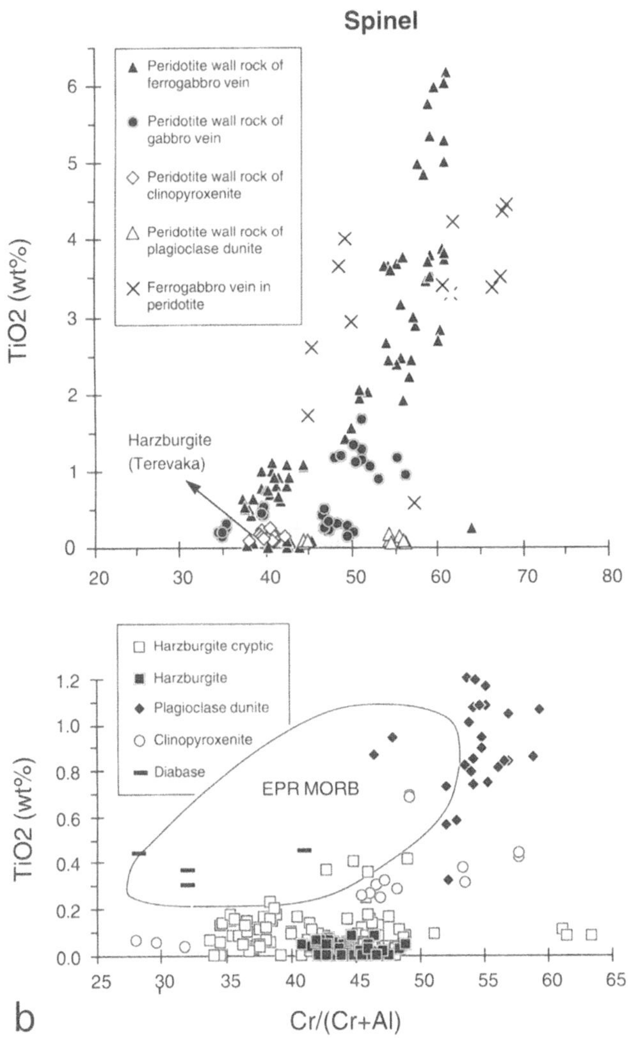

Figure 2 continued. Compositional variations of spinels from the Terevaka transform fault samples: (b) variation diagrams of TiO_2 (weight %) versus Cr#. All symbols as in (a). Note extreme TiO_2 enrichment of spinels in peridotite wall rock adjacent to intrusive ferrogabbro.

We have also identified regular rims along lithological contacts (Plates 1B, 2B, 2E, 3C), and granoblastic, fine-grained mottled zones located between several magmatic veins (black dunitic part in Plate 1F and 1G), and these are treated separately.

The mineralogical associations and crosscutting relationships of the various veins allow the identification and relative chronology of successive magmatic generations in the harzburgite from the Terevaka transform fault, from the earliest to the latest intrusive event (Table 2). They are divided as follows: (1) clinopyroxenite veins and patches in harzburgite; (2) massive plagioclase-dunites and dunite rims in harzburgite; (3) primitive fine-grained olivine-bearing microgabbro dikelets; (4) mylonitic gabbro and medium-grained undeformed gabbro veins; (5) coarse-grained vein networks and pods of ferrogabbro; (6) mylonitic fine-grained ferrogabbro veins (kaersutite-, apatite-, zircon-bearing) often injected into pre-existent gabbroic veins. Some overlap exist between types 3 and 4. The metamorphic events superimposed on all listed intrusion types and host peridotites include amphibole veining and pseudomorphism by talc, Mg-rich colourless amphibole and green amphibole, followed by low-temperature serpentinization veins and associated pervasive metamorphism and, finally, sea-floor carbonate and argile precipitation.

All magmatic vein types show examples of diffuse, olivine-rich aureoles or massive, plagioclase-free peridotitic and dunitic clusters or rims developed in the peridotite wall rock (Plate 2E). Many samples show crosscutting relationships between the various intrusive types. In sample 83DS-M163, for example, a clinopyroxenite vein parallel to the host peridotite foliation is crosscut at a high angle by an amphibolitized gabbro vein (right hand part of Plate 1A). Dunitic harzburgite samples 83DS-M28 and M29 have multiple generations of gabbroic veins, cutting through centimeter-scale peridotitic clusters made of abundant equigranular, fine-grained recrystallized olivine and orthopyroxene grains, and of minor coarse-grained deformed orthopyroxene porphyroclasts surrounded by small olivine granoblasts (Plates 2E, 3G). Basaltic dike sample 83DS-M13 (not shown) shows a quenched margin suggesting rapid cooling against a coarse-grained porphyroclastic amphibolitized gabbro with calcic plagioclase (An81).

Mineral chemistry

Microprobe procedures

Mineral compositions for all of the data set were determined using a Camebax SX50 microprobe at IFREMER (Brest). Our routine analytical conditions were 6 seconds of counting, 15 nA sample current, and 15 kV accelerating potential. Additional high-precision analyses of NiO, TiO_2, Cr_2O_3 in olivine, clinopyroxene and orthopyroxene were performed with 25 seconds of counting, 80 nA sample current and 25 kV accelerating potential. These operating conditions are similar to those used by Cannat et al. (1990) for peridotites of the Garrett transform fault. In order to evaluate the nature of chemical exchanges, minerals from host peridotites and intrusive gabbros were systematically analyzed at proximity and away from the contact between the two lithologies. Mineral compositions representing averaged analyses for each of the rock types studied are listed in Tables 3 to 8.

Spinel

The spinel chemistry is particularly useful for deciphering the petrogenesis of peridotites since this mineral is resistant to serpentinization, always present in at least trace amounts, and sensible to solid-liquid reactions. Several distinct compositional groups are identified according to rock type (Table 3; Fig. 2a, b). Harzburgites free of impregnation and intrusion contain coarse-grained, interstitial to poikilitic brown spinel with Cr#=40-50, Mg#=60-70, and TiO_2<0.1%. The spinels from the host harzburgite located several cm away from any sharp gabbroic contact are compositionally identical and are not affected by the melt brought by the veins. A subset called cryptic harzburgite designates the analyzed zones located within 1 cm of any intrusion but away from the immediate wall rock contact. These spinels extend to distinctly lower Cr#=35 and higher Mg#=75 and TiO_2 up to 0.4% (Fig. 2a). Plagioclase dunites have black spinel with higher Cr#=55 and lower Mg#=45-55 than harzburgites. The spinels from peridotite wall rock in contact with intrusive gabbros, and those within clinopyroxenite veins generally show intermediate compositions between the host harzburgite and peridotite wall rock of intrusive ferrogabbros. The trend extending toward low Mg#, high Cr# (Fig. 2a) and very TiO_2-rich (Fig. 2b) compositions are spinels analyzed along the wall rock margins of ferrogabbroic veins but located within the host peridotite. This Fe-enrichment trend is limited by the disappearance of spinel as stable phase in the adjacent ferrogabbro, where abundant Fe-Ti oxide occurs. These spinels are the richest in TiO_2 found in the ocean basins, and reflect the extreme chemical disequilibrium between the ferrogabbro and the peridotite (Fig. 2b). Calculated Fe3# values are low (Fe3#<5) except for Ti-rich spinel from the peridotite wall rock of sample 83DS-24 adjacent to an intrusive ferrogabbro (Fe3#=18.4).

Olivine

Analyses of olivine are subdivided in terms of lithology (Table 4; Fig. 3a, b). Harzburgite free of any intrusion has a refractory and homogeneous olivine composition (on average Fo=91, NiO=0.38-0.40%, MnO<0.25%). Dunite veinlets have olivine with similar Fo and Mn contents as in harzburgite, but with lower Ni values (NiO=0.24%, sample 83DS-26, -42; Fig. 3a). On the other hand, plagioclase-dunites have lower Fo and NiO content than the harzburgites (Fig. 3a). The wall rock peridotite of the intrusive gabbro group has compositions intermediate between harzburgite and pla-

Table 4. Microprobe analyses of olivine in mafic and ultramafic rocks from the Terevaka transform fault and the Pito Deep. Rock types include PER: host harzburgite, PLDU: plag-dunite, CPXN: clinopyroxenite, DUN: dunite, G: gabbro, FG: ferrogabbro, LTR: leucotroctolite, MIGO: olivine microgabbro, v: vein in peridotite, z: wall rock reaction zone adjacent to contact, c: cryptic zone (≤ 1cm from intrusion), g: granoblast.
avg=average number of microprobe analyses (Ni=with long counting time), n.d.=not determined, Fo=100*Mg/(Mg+Fe).

Sample	*11DS*	*18DS-5*	*26*	*26*	*26R*	*28*	*42*	*42*	*42*	*42*	*M11*	*M12*	*M12*	*M12*	*M14*
Rock	*LTR*	*G*	*PERc*	*DUNv*	*PERz*	*PER*	*MIGOv*	*DUNvg*	*PERz*	*PERz*	*PLDU*	*PERc*	*PERzg*	*FGvz*	*PERz*
avg	*6 Ni*	*7 (4Ni)*	*4 (3Ni)*	*3 Ni*	*7 (3Ni)*	*14 (5Ni)*	*2 Ni*	*3 Ni*	*3 (2Ni)*	*3 Ni*	*7 (6Ni)*	*8 (3Ni)*	*5 (4Ni)*	*9*	*4 (3Ni)*
SiO_2	40.53	38.08	41.24	41.04	40.87	41.41	38.53	40.53	40.01	40.74	40.69	41.30	40.18	32.15	40.75
FeO	11.93	24.46	8.71	8.59	10.21	9.22	22.87	9.39	15.55	10.80	11.40	9.13	15.66	56.89	9.72
MnO	0.19	0.36	0.10	0.14	0.12	0.14	0.40	0.14	0.28	0.19	0.16	0.13	0.27	1.34	0.12
MgO	47.29	36.64	50.38	51.11	49.05	50.44	39.32	49.90	44.87	49.22	48.82	50.56	45.06	11.27	49.79
NiO	0.17	0.10	0.38	0.24	0.34	0.38	0.15	0.24	0.30	0.31	0.26	0.39	0.34	n.d.	0.36
CaO	0.03	0.03	0.00	0.00	0.01	0.02	0.22	0.01	0.01	0.01	0.04	0.02	0.02	0.11	0.02
Total	100.26	99.73	100.74	101.14	100.45	101.39	101.52	100.24	101.01	101.32	101.37	101.37	101.46	101.80	100.70
O=4															
Si	1.000	1.003	0.999	0.990	0.999	0.999	0.991	0.991	0.997	0.992	0.992	0.996	0.998	0.989	0.993
Fe	0.247	0.540	0.176	0.173	0.209	0.186	0.492	0.192	0.325	0.220	0.232	0.184	0.325	1.464	0.198
Mn	0.003	0.010	0.002	0.003	0.002	0.000	0.009	0.003	0.006	0.004	0.003	0.003	0.006	0.035	0.003
Mg	1.740	1.441	1.818	1.838	1.787	1.812	1.508	1.818	1.667	1.785	1.774	1.816	1.667	0.517	1.808
Ni	0.000	0.000	0.006	0.005	0.003	0.004	0.003	0.005	0.004	0.006	0.004	0.003	0.005	0.000	0.005
Ca	0.000	0.000	0.000	0.000	0.000	0.000	0.006	0.000	0.000	0.000	0.001	0.000	0.000	0.004	0.001
Fo	87.61	72.76	91.16	91.38	89.54	90.70	75.40	90.45	83.71	89.04	88.41	90.79	83.67	26.08	90.13

Table 4 continued. Microprobe analyses of olivine in mafic and ultramafic rocks from the Terevaka transform fault and the Pito Deep.

Sample	*M14*	*M14*	*M18*	*M21*	*M21*	*M21*	*M21*	*M28*	*M28*	*M28*	*M28*	*M28*	*M28*	*M33*	*M163*
Rock	*PERz*	*PERc*	*PLDU*	*FGvz*	*DUNz*	*PERcg*	*PERc*	*CPXN*	*PERc*	*PERc*	*PERz*	*PERzg*	*PERzg*	*CPXN*	*PERc*
avg	*4 (3Ni)*	*4 (2Ni)*	*9*	*3*	*4*	*2*	*2*	*1*	*6*	*2*	*7*	*3*	*3*	*3*	*4*
SiO_2	40.75	41.42	41.55	32.58	39.05	40.95	41.19	41.58	41.38	41.68	41.22	40.45	37.11	41.63	41.32
FeO	9.72	8.95	11.44	52.96	20.02	10.79	8.64	9.03	8.84	9.11	10.33	15.10	32.25	11.00	9.16
MnO	0.12	0.18	0.13	1.24	0.30	0.22	0.11	0.08	0.14	0.20	0.13	0.26	0.64	0.18	0.11
MgO	49.79	50.75	48.58	13.55	41.57	49.08	50.58	50.72	50.54	49.48	48.74	44.81	30.91	48.85	49.87
NiO	0.36	0.37	n.d.	n.d.	n.d.	n.d.	n.d.	n.d.	n.d.	n.d.	n.d.	n.d.	n.d.	n.d.	n.d.
CaO	0.02	0.01	0.02	0.06	0.02	0.00	0.00	0.05	0.01	0.01	0.01	0.01	0.04	0.04	0.02
Total	100.70	101.52	101.80	100.44	101.01	101.17	100.54	101.51	101.14	100.54	100.50	100.72	101.07	101.75	100.55
O=4															
Si	0.993	0.996	1.005	0.996	0.994	0.996	0.998	0.999	0.998	1.011	1.006	1.007	1.003	1.006	1.003
Fe	0.198	0.180	0.231	1.354	0.427	0.220	0.175	0.181	0.178	0.185	0.211	0.315	0.729	0.222	0.186
Mn	0.003	0.004	0.003	0.032	0.006	0.005	0.003	0.002	0.003	0.005	0.003	0.006	0.015	0.004	0.002
Mg	1.808	1.820	1.752	0.618	1.577	1.779	1.827	1.816	1.817	1.788	1.772	1.663	1.245	1.759	1.804
Ni	0.005	0.004	0.000	0.000	0.000	0.000	0.000	0.000	0.004	0.000	0.000	0.000	0.000	0.000	0.000
Ca	0.001	0.000	0.001	0.002	0.001	0.000	0.000	0.001	0.000	0.000	0.000	0.000	0.001	0.001	0.001
Fo	90.13	91.00	88.33	30.82	78.45	88.82	91.14	90.85	91.00	90.64	89.36	84.10	63.06	88.78	90.65

gioclase-dunite (Fig. 3a, b). The two groups of peridotite reaction zones caused by gabbro and ferrogabbro intrusions have distinct Fo and MnO contents for similar high NiO content (Fig. 3a, b). In the Pito Deep, leucotroctolites are relatively Mg- and Ni-rich (Fo86-88, NiO=0.16-0.2%) as compared to massive gabbro compositions (Fig. 3a, b; Fo73-78, NiO=0.06-0.12%).

Orthopyroxene

The harzburgite contains medium to coarse grained enstatites which are plastically deformed with similar Mg-rich compositions in core and rim (Table 5; Fig. 4b). Zoned enstatite grains occur in peridotite wall rock near intrusive gabbros. They are usually granular and undeformed, and have average core compositions almost similar to enstatite in harzburgite (Fig. 4c), whilst their rims have reequilibrated, presumably with the infiltrating evolved liquid, to lower Mg#, Al_2O_3 and Cr_2O_3 contents (on average Mg#=86.5, Al_2O_3=2.6%, Cr_2O_3=0.68%; Fig. 4b, 4d). The small orthopyroxene granoblasts corroding these orthopyroxenes are even more evolved

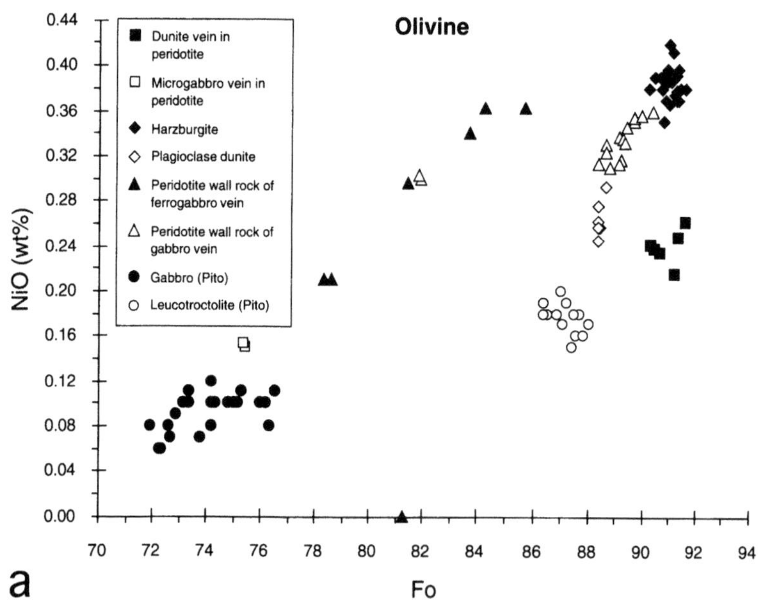

Figure 3. Compositional variations of olivines in samples from the Pito Deep and Terevaka transform fault: (a) variation diagram of NiO (weight %) versus Fo content. Massive gabbro (18DS) and leucotroctolite (11DS) samples from Pito Deep, all other samples from Terevaka. Dunite vein is located near intrusive gabbro contact but within peridotite wall rock. All NiO analyses measured with 30 seconds counting time. See text for details.

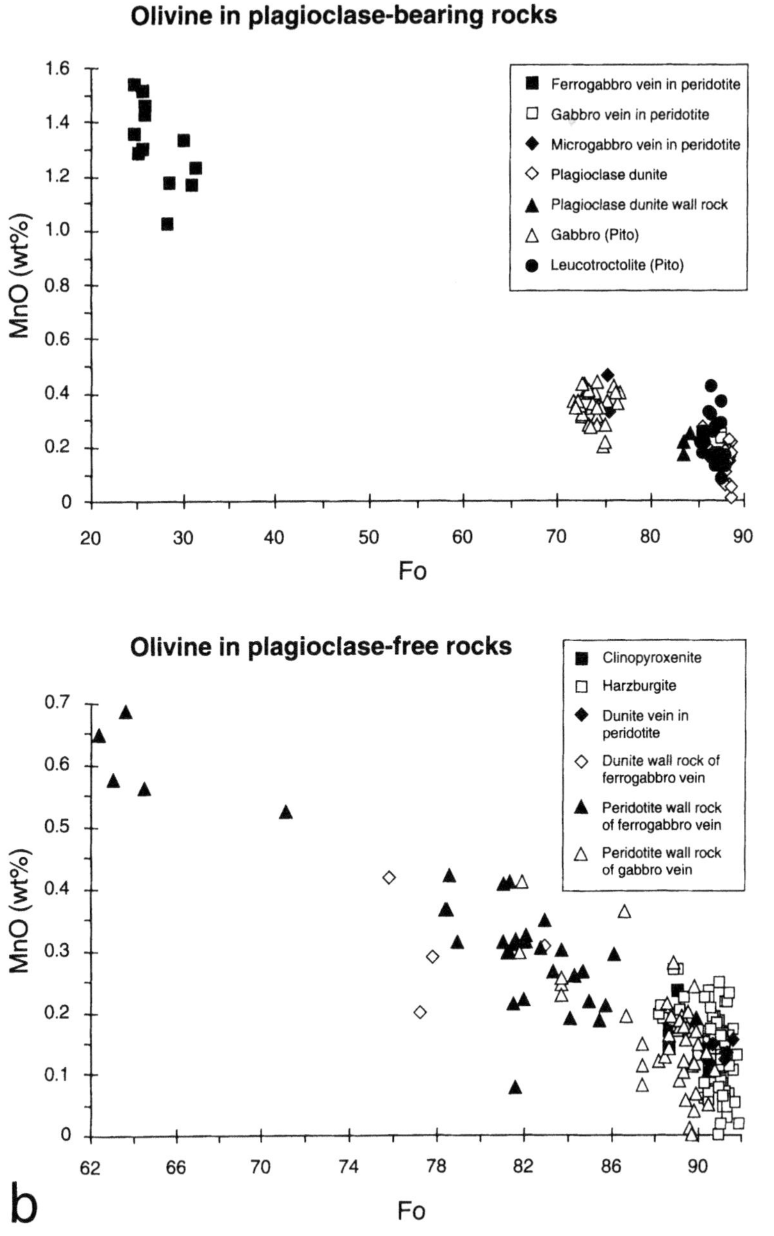

Figure 3. continued. Compositional variations of olivines in samples from the Pito Deep and Terevaka transform fault: (b) variation diagrams of MnO (weight %) versus Fo content. Upper and lower diagrams show olivine from plagioclase-bearing and plagioclase-free rocks, respectively. Note decrease in Fo content concurrent with MnO increase, both in ferrogabbro intrusion (Fo30) and in adjacent wall-rock peridotite reaction zone (Fo86-62).

Table 5. Microprobe analyses of orthopyroxene in mafic and ultramafic rocks from the Terevaka transform fault and the Pito Deep. Rock types include PER: host harzburgite, PLDU: plagioclase dunite, CPXN: clinopyroxenite, DUN: dunite, G: gabbro, LTR: leucotroctolite, MYG: mylonitic gabbro, MIGO: olivine microgabbro, FG: ferrogabbro, v: vein in peridotite, z: wall rock reaction zone adjacent to contact, c: cryptic zone (≤ 1cm from intrusion), g: granoblast. avg=average number of microprobe analyses, Mg#=100*Mg/(Mg+Fe).

Sample	*11DS-2*	*18DS-5*	*26*	*26*	*28*	*42*	*42*	*42*	*M11*	*M12*	*M14*	*M14*	*M21*
Rock	*LTR*	*G*	*PERc*	*PERcg*	*PER*	*MIGOv*	*PERc*	*PERzg*	*PLDU*	*PERzg*	*MYGv*	*PERc*	*PERc*
avg	*4*	*2*	*19*	*4*	*12*	*1*	*8*	*5*	*4*	*14*	*4*	*11*	*3*
SiO_2	56.49	54.36	56.66	57.46	57.00	56.22	56.41	56.80	55.81	56.60	56.16	56.80	56.62
TiO_2	0.41	0.30	0.02	0.01	0.02	0.49	0.05	0.05	0.18	0.02	0.55	0.01	0.02
Al_2O_3	1.46	1.11	2.19	1.84	2.43	1.24	2.74	2.46	2.16	2.52	1.72	2.49	2.59
Cr_2O_3	0.33	0.05	0.58	0.39	0.64	0.13	0.79	0.71	0.65	0.68	0.08	0.77	0.65
FeO	9.43	14.78	5.94	5.59	6.06	9.67	6.11	7.12	7.04	6.20	9.63	5.95	5.38
MnO	0.30	0.44	0.11	0.13	0.13	0.24	0.10	0.18	0.16	0.13	0.25	0.13	0.07
MgO	32.24	26.97	33.65	34.76	33.14	31.89	32.78	32.77	33.33	33.67	31.98	34.19	33.90
CaO	0.64	0.97	1.08	0.76	1.47	0.99	1.62	1.12	1.60	1.41	1.00	1.45	1.42
Na_2O	0.01	0.04	0.05	0.02	0.01	0.01	0.04	0.04	0.01	0.02	0.03	0.01	0.04
Total	101.33	99.02	100.31	100.96	100.94	100.92	100.65	101.27	101.01	101.30	101.43	101.84	100.70
O=6													
Si	1.954	1.974	1.950	1.958	1.951	1.956	1.939	1.947	1.925	1.935	1.944	1.930	1.938
Ti	0.011	0.008	0.000	0.000	0.000	0.013	0.001	0.001	0.004	0.000	0.014	0.000	0.000
Al IV	0.046	0.026	0.050	0.042	0.049	0.044	0.061	0.053	0.075	0.066	0.056	0.070	0.062
Al VI	0.014	0.022	0.039	0.032	0.050	0.007	0.050	0.046	0.013	0.036	0.014	0.030	0.043
Cr	0.009	0.002	0.016	0.011	0.018	0.004	0.021	0.019	0.018	0.018	0.002	0.021	0.018
Fe	0.273	0.449	0.171	0.159	0.174	0.281	0.176	0.204	0.203	0.177	0.279	0.169	0.154
Mn	0.009	0.013	0.003	0.004	0.004	0.007	0.003	0.005	0.005	0.004	0.007	0.003	0.002
Mg	1.662	1.460	1.726	1.765	1.691	1.654	1.680	1.674	1.713	1.715	1.650	1.732	1.729
Ca	0.024	0.038	0.040	0.028	0.054	0.037	0.059	0.041	0.059	0.052	0.037	0.053	0.052
Na	0.000	0.003	0.004	0.001	0.000	0.001	0.003	0.003	0.000	0.002	0.002	0.000	0.003
Mg#	85.90	76.48	91.00	91.73	90.69	85.46	90.53	89.13	89.40	90.63	85.55	91.11	91.82
Wo	1.21	1.95	2.09	1.43	2.81	1.88	3.10	2.16	3.01	2.66	1.89	2.71	2.69
En	84.86	74.99	89.09	90.41	88.14	83.87	87.72	87.22	86.72	88.22	83.94	88.64	89.35
Fs	13.93	23.06	8.82	8.16	9.05	14.25	9.18	10.63	10.28	9.12	14.17	8.65	7.96

Table 5 continued. Microprobe analyses of orthopyroxene in mafic and ultramafic rocks from the Terevaka transform fault and the Pito Deep.

Sample	*M21*	*M21*	*M28*	*M28*	*M28*	*M28*	*M28*	*M28*	*M33*	*M33*	*M163*	*M163*	*M163*
Rock	*PERzg*	*DUNz*	*PERc*	*PERz*	*CPXN*	*PERz*	*PERzg*	*FGvg*	*Gz*	*CPXN*	*PERc*	*PERzg*	*CPXN*
avg	*2*	*1*	*3*	*3*	*4*	*8*	*8*	*2*	*4*	*3*	*5*	*2*	*8*
SiO_2	57.14	55.12	56.34	55.47	57.20	56.05	55.09	53.95	54.99	55.38	56.63	56.76	57.18
TiO_2	0.04	0.18	0.08	0.13	0.08	0.09	0.09	0.25	0.42	0.06	0.01	0.04	0.16
Al_2O_3	1.75	0.53	3.23	3.25	2.53	2.82	1.93	0.54	1.46	3.52	2.71	2.43	1.43
Cr_2O_3	0.29	0.10	0.77	0.87	0.58	0.85	0.48	0.09	0.10	0.89	0.74	0.54	0.46
FeO	6.74	13.42	6.11	5.28	6.29	6.83	13.10	20.43	15.63	8.63	5.79	5.73	6.65
MnO	0.28	0.36	0.12	0.17	0.19	0.16	0.41	0.56	0.39	0.18	0.14	0.08	0.14
MgO	34.12	28.34	34.30	32.36	34.14	31.75	28.44	23.38	27.61	31.12	33.44	33.74	33.97
CaO	0.85	1.03	1.01	3.55	0.79	2.18	1.19	1.24	1.32	2.66	1.20	0.76	0.83
Na_2O	0.02	0.00	0.02	0.05	0.00	0.07	0.06	0.00	0.04	0.05	0.04	0.00	0.01
Total	101.25	99.08	101.98	101.14	101.82	100.81	100.80	100.46	101.98	102.49	100.71	100.08	100.86
O=6													
Si	1.954	1.987	1.911	1.906	1.940	1.935	1.952	1.985	1.948	1.901	1.941	1.952	1.961
Ti	0.001	0.005	0.002	0.003	0.002	0.002	0.003	0.007	0.011	0.002	0.000	0.001	0.004
Al IV	0.047	0.013	0.089	0.094	0.060	0.066	0.048	0.015	0.052	0.099	0.059	0.048	0.039
Al VI	0.024	0.009	0.041	0.038	0.041	0.049	0.032	0.008	0.009	0.044	0.050	0.051	0.019
Cr	0.008	0.003	0.020	0.023	0.016	0.023	0.013	0.003	0.003	0.024	0.020	0.015	0.013
Fe	0.193	0.405	0.173	0.152	0.178	0.198	0.392	0.629	0.463	0.248	0.166	0.165	0.191
Mn	0.008	0.011	0.003	0.005	0.006	0.005	0.012	0.018	0.012	0.005	0.004	0.002	0.004
Mg	1.739	1.523	1.734	1.658	1.726	1.633	1.498	1.283	1.458	1.592	1.708	1.730	1.737
Ca	0.031	0.040	0.037	0.131	0.029	0.081	0.045	0.049	0.050	0.098	0.044	0.028	0.031
Na	0.001	0.000	0.002	0.003	0.000	0.005	0.004	0.000	0.002	0.003	0.003	0.000	0.001
Mg#	90.00	78.99	90.91	91.61	90.64	89.13	79.24	67.11	75.90	86.56	91.15	91.29	90.11
Wo	1.58	2.03	1.89	6.73	1.48	4.24	2.35	2.48	2.54	5.07	2.31	1.45	1.57
En	88.58	77.39	89.20	85.44	89.30	85.40	77.42	65.45	73.97	82.13	89.05	89.97	88.69
Fs	9.84	20.58	8.92	7.83	9.22	10.36	20.23	32.08	23.50	12.80	8.65	8.58	9.73

Table 6. Microprobe analyses of clinopyroxene in mafic and ultramafic rocks from the Terevaka transform fault and the Pito Deep. Rock types include PER: host harzburgite, PLDU: plagioclase dunite, CPXN: clinopyroxenite, G: gabbro, MG: metagabbro, LTR: leucotroctolite, DIA: diabase, MYG: mylonitic gabbro, MIGO: olivine microgabbro, FG: ferrogabbro, v: vein in peridotite, m: middle of vein, z: wall rock reaction zone adjacent to contact, c: cryptic zone (≤ 1cm from intrusion), g: granoblast. avg=average number of microprobe analyses, Mg#=100*Mg/(Mg+Fe).

Sample	*11DS*	*18DS-5*	*26*	*28*	*28*	*28*	*32*	*34*	*42*	*42*	*42*	*M11*	*M12*	*M12*	*M14*
Rock	*LTR*	*G*	*MG*	*MYGv*	*PERc*	*PER*	*G*	*DIA*	*MIGOvm*	*MIGOv*	*MIGOvz*	*PLDU*	*FGv*	*FGvz*	*Gv*
avg	*6*	*10*	*6*	*11*	*3*	*10*	*10*	*9*	*3*	*4*	*4*	*13*	*12*	*13*	*5*
SiO_2	51.03	51.96	52.37	53.36	53.38	53.58	52.97	52.30	52.98	52.24	52.43	52.71	50.23	50.97	50.99
TiO_2	0.84	0.62	1.08	0.42	0.04	0.01	0.58	0.68	0.76	1.26	1.31	0.41	0.40	0.40	1.69
Al_2O_3	3.63	2.16	2.85	1.99	2.16	2.70	3.22	2.44	2.27	3.08	3.51	3.49	0.90	0.92	4.20
Cr_2O_3	0.94	0.22	0.15	0.20	1.02	0.87	0.31	0.16	0.45	0.32	0.35	1.29	0.03	0.00	1.00
FeO	3.64	6.65	5.45	3.24	2.46	2.36	5.63	10.21	5.53	4.61	2.98	3.40	21.05	20.05	3.72
MnO	0.10	0.18	0.16	0.10	0.09	0.09	0.18	0.27	0.17	0.17	0.10	0.12	0.69	0.68	0.12
MgO	16.51	15.23	16.79	17.78	17.40	18.04	16.99	17.67	16.30	16.73	16.89	18.06	8.11	9.62	16.10
CaO	22.43	21.90	21.02	21.77	22.24	23.27	21.06	16.93	22.06	21.55	22.66	20.88	18.59	17.91	22.21
Na_2O	0.24	0.34	0.43	0.34	0.45	0.05	0.51	0.23	0.41	0.47	0.50	0.37	0.37	0.34	0.57
Total	99.38	99.29	100.31	99.19	99.24	100.98	101.44	100.90	100.94	100.44	100.75	100.72	100.40	100.91	100.61
O=6															
Si	1.880	1.935	1.913	1.952	1.951	1.926	1.913	1.918	1.929	1.904	1.895	1.902	1.966	1.968	1.857
Ti	0.022	0.019	0.030	0.011	0.001	0.000	0.016	0.019	0.021	0.034	0.036	0.011	0.012	0.012	0.046
Al IV	0.120	0.065	0.087	0.048	0.049	0.074	0.087	0.082	0.071	0.096	0.105	0.098	0.034	0.032	0.143
Al VI	0.038	0.032	0.036	0.038	0.044	0.040	0.050	0.024	0.027	0.036	0.045	0.050	0.008	0.010	0.038
Cr	0.028	0.008	0.005	0.006	0.029	0.024	0.009	0.005	0.013	0.009	0.010	0.037	0.001	0.000	0.029
Fe	0.113	0.208	0.167	0.099	0.075	0.071	0.170	0.313	0.168	0.140	0.090	0.102	0.689	0.648	0.113
Mn	0.000	0.009	0.005	0.003	0.002	0.003	0.005	0.008	0.005	0.005	0.003	0.003	0.023	0.022	0.004
Mg	0.905	0.847	0.914	0.969	0.948	0.967	0.915	0.965	0.885	0.909	0.910	0.971	0.473	0.553	0.874
Ca	0.887	0.874	0.823	0.853	0.871	0.897	0.815	0.666	0.861	0.841	0.878	0.807	0.780	0.741	0.867
Na	0.018	0.025	0.030	0.024	0.031	0.003	0.035	0.016	0.029	0.034	0.035	0.026	0.028	0.026	0.040
Mg#	88.98	80.38	84.58	90.73	92.64	93.17	84.32	75.94	84.02	86.61	90.99	90.49	40.74	46.10	88.57
Wo	46.54	45.30	43.23	44.39	45.97	46.35	42.88	34.29	44.97	44.49	46.75	42.94	40.14	38.15	46.77
En	47.51	43.91	48.01	50.44	50.05	49.98	48.16	49.66	46.24	48.08	48.46	51.62	24.37	28.48	47.13
Fs	5.95	10.79	8.76	5.17	3.98	3.67	8.96	16.05	8.79	7.42	4.79	5.44	35.49	33.37	6.10

Table 6 continued. Microprobe analyses of clinopyroxene in mafic and ultramafic rocks from the Terevaka transform fault and the Pito Deep.

Sample	*M14*	*M18*	*M18*	*M21*	*M21*	*M28*	*M28*	*M28*	*M28*	*M28*	*M33*	*M33*	*M33*	*M163*	*M163*
Rock	*MYGv*	*PERz*	*PLDUv*	*FGv*	*PERc*	*CPXNg*	*FGvg*	*FGv*	*Gvz*	*Gvz*	*Gz*	*CPXNz*	*CPXN*	*CPXN*	*PERc*
avg	*7*	*5*	*12*	*7*	*1*	*9*	*4*	*3*	*3*	*2*	*11*	*4*	*8*	*19*	*9*
SiO_2	52.67	54.13	53.49	51.25	52.84	53.02	52.14	51.15	53.30	52.92	52.68	53.00	52.94	53.33	53.42
TiO_2	1.27	0.27	0.86	0.41	0.14	0.25	0.38	0.42	0.59	0.64	0.84	0.20	0.20	0.37	0.02
Al_2O_3	3.21	2.40	2.78	1.14	3.41	3.94	1.10	0.90	2.22	2.85	2.84	3.66	3.92	2.51	2.77
Cr_2O_3	0.39	1.13	0.44	0.05	1.12	1.07	0.08	0.03	0.95	1.05	0.45	1.08	1.31	0.89	0.86
FeO	3.34	2.73	3.01	17.11	2.44	2.47	12.34	18.40	2.43	2.77	5.87	4.28	3.04	2.52	2.16
MnO	0.14	0.09	0.11	0.51	0.03	0.08	0.50	0.57	0.07	0.00	0.20	0.15	0.10	0.14	0.09
MgO	16.96	17.68	18.32	11.58	18.05	17.63	13.05	9.89	17.24	17.91	16.30	17.44	17.22	17.24	17.48
CaO	22.48	22.95	22.11	17.87	23.03	22.56	20.13	18.37	22.92	21.13	21.74	21.14	22.20	22.84	23.01
Na_2O	0.41	0.47	0.41	0.40	0.42	0.39	0.35	0.39	0.45	0.43	0.39	0.41	0.42	0.34	0.33
Total	100.89	101.85	101.53	100.36	101.50	101.43	100.07	100.14	100.19	99.69	101.32	101.36	101.35	100.18	100.17
O=6															
Si	1.903	1.933	1.914	1.961	1.894	1.898	1.968	1.977	1.934	1.923	1.913	1.907	1.901	1.934	1.934
Ti	0.035	0.007	0.023	0.012	0.004	0.007	0.011	0.012	0.016	0.018	0.023	0.006	0.006	0.010	0.000
Al IV	0.097	0.067	0.085	0.039	0.106	0.102	0.033	0.023	0.066	0.077	0.087	0.094	0.100	0.066	0.066
Al VI	0.039	0.034	0.032	0.013	0.038	0.064	0.017	0.018	0.029	0.045	0.034	0.062	0.065	0.042	0.052
Cr	0.011	0.032	0.012	0.002	0.032	0.030	0.003	0.001	0.027	0.030	0.013	0.031	0.037	0.025	0.025
Fe	0.101	0.082	0.090	0.548	0.073	0.074	0.390	0.595	0.074	0.084	0.178	0.129	0.091	0.076	0.065
Mn	0.004	0.003	0.003	0.017	0.001	0.002	0.016	0.018	0.002	0.000	0.006	0.005	0.003	0.004	0.003
Mg	0.913	0.941	0.977	0.659	0.964	0.940	0.734	0.569	0.933	0.970	0.882	0.935	0.921	0.932	0.943
Ca	0.870	0.878	0.847	0.733	0.884	0.865	0.814	0.761	0.891	0.823	0.846	0.815	0.854	0.887	0.893
Na	0.029	0.032	0.029	0.030	0.029	0.027	0.026	0.029	0.032	0.030	0.027	0.029	0.029	0.024	0.023
Mg#	90.10	92.03	91.57	54.64	92.96	92.77	65.36	48.93	92.65	92.06	83.19	87.89	91.00	92.44	93.53
Wo	46.17	46.19	44.29	37.76	46.02	46.05	42.01	39.52	46.95	43.86	44.37	43.38	45.76	46.82	46.96
En	48.46	49.51	51.01	33.97	50.18	50.03	37.88	29.58	49.15	51.67	46.27	49.76	49.36	49.16	49.61
Fs	5.36	4.29	4.71	28.27	3.80	3.92	20.11	30.89	3.90	4.47	9.36	6.87	4.88	4.02	3.43

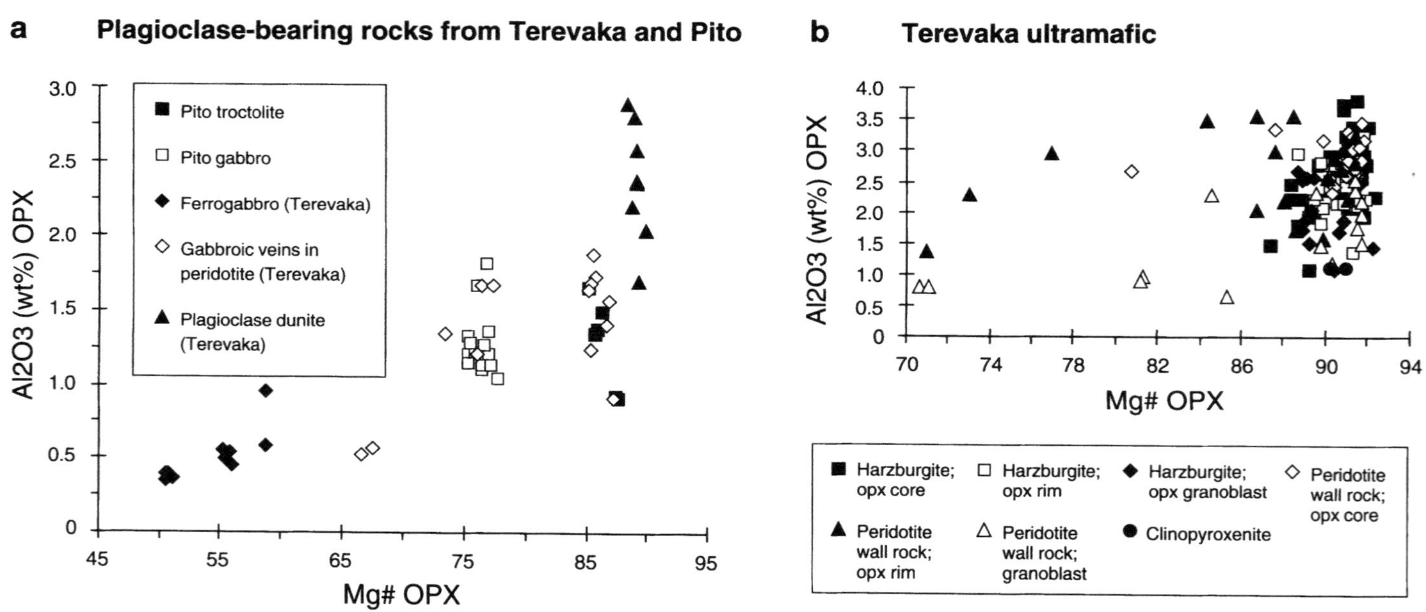

Figure 4. Orthopyroxene Al_2O_3 (weight %) versus Mg# in samples from the Terevaka transform fault and gabbroic rocks from Pito Deep: (a) orthopyroxenes in plagioclase-bearing rocks; (b) orthopyroxenes in ultramafic rocks.

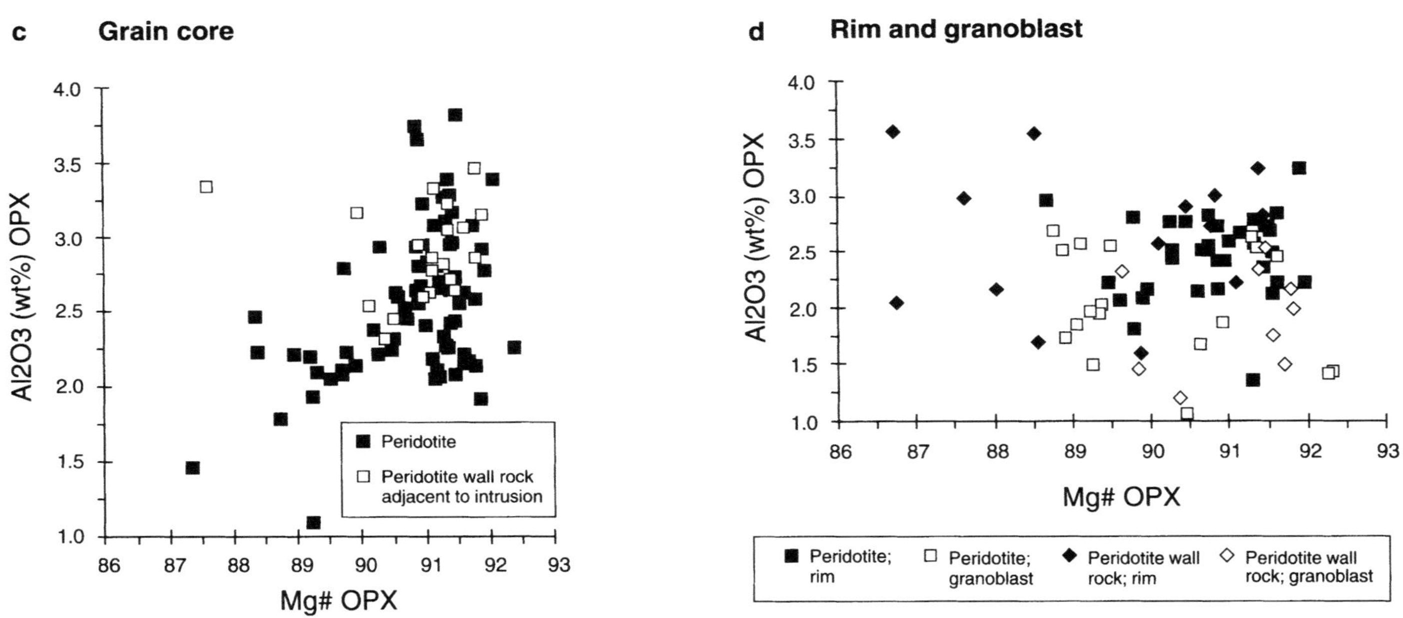

Figure 4 continued. Orthopyroxene Al_2O_3 (weight %) versus Mg# in samples from the Terevaka transform fault and gabbroic rocks from Pito Deep: (c, d) details of (4b) with orthopyroxene cores and rims Mg# > 86 from peridotites. Note that, on average, rims of orthopyroxene from harzburgite have lower Mg# and Al_2O_3 contents than the cores of the same grains.

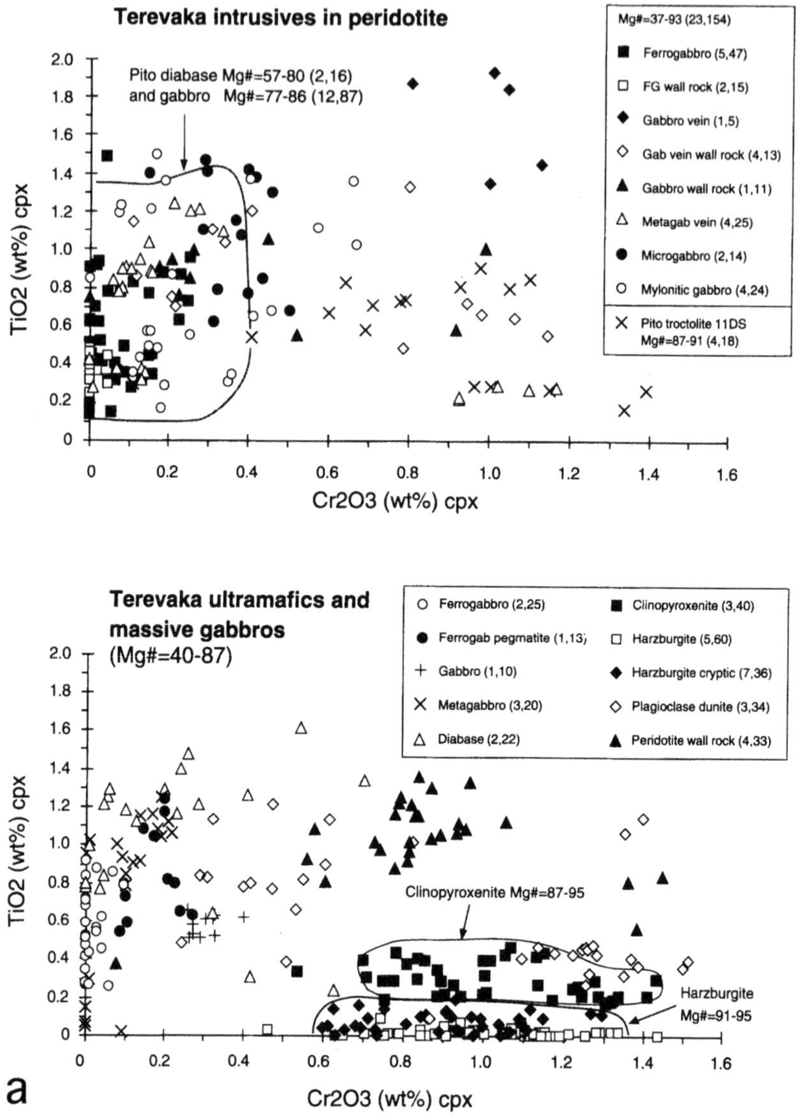

Figure 5. Compositional variations of clinopyroxenes in samples from the Terevaka transform fault and gabbroic rocks from Pito Deep. The two numbers in parentheses indicated for each rock type (legend) denote number of samples and number of analyses, respectively. (a) Variation diagrams of TiO_2 (weight %) versus Cr_2O_3(weight %). Upper diagram includes analyses in all of the various gabbros intruding the peridotites from Terevaka. Field of high Ti, low Cr (surrounded by line) refers to unplotted analyses of massive gabbro and dolerite from the Pito Deep with their range of Mg#. Most clinopyroxenes from intrusive gabbros plot inside this compositional field. Lower diagram shows clinopyroxenes in ultramafics and in massive gabbros from Terevaka. Clinopyroxenites and harzburgites have diopside with low Ti, high Cr and Mg contents.

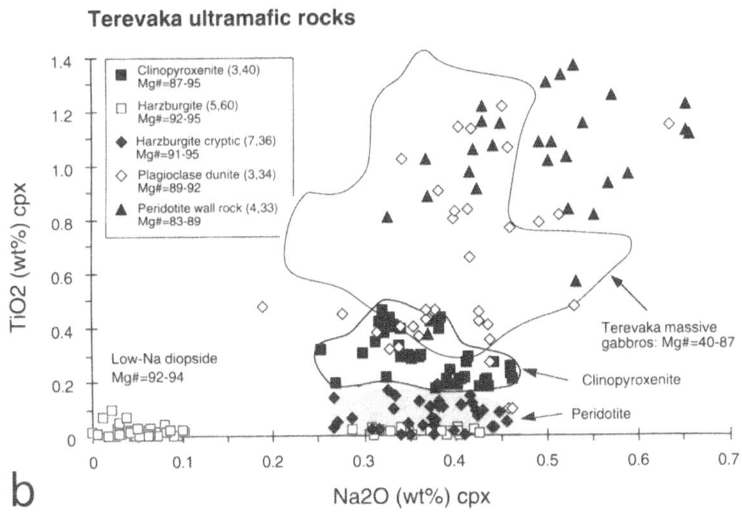

Figure 5 continued. Compositional variations of clinopyroxenes in samples from the Terevaka transform fault and gabbroic rocks from Pito Deep:.(b) variation diagram of TiO_2 (weight %) versus Na_2O (weight %). Symbols as in (a), lower diagram. Note the two diopside subgroups in harzburgite defined by their different Na_2O content for both extremely low TiO_2 values. See text for discussion.

(on average Mg#=86.2, Al_2O_3=1.6%, Cr_2O_3=0.39%; Fig. 4b, 4d). When compared to harzburgite, clinopyroxenite has minor amounts of small enstatite granoblasts characterized by lower Al_2O_3, Cr_2O_3 and slightly higher TiO_2 values, whilst orthopyroxene in plagioclase-dunite has slightly lower Mg# and higher TiO_2 values (Fig. 4a). Gabbroic veins and massive gabbros from Terevaka, and leucotroctolites and gabbros from Pito Deep have exsolved or interstitial orthopyroxene with comparatively low Mg#, Al_2O_3, Cr_2O_3, and high TiO_2 values (Table 5).

Clinopyroxene

Harzburgite has small granular, interstitial or exsolved diopside grains which have on average the lowest Al_2O_3 (2.8%), TiO_2 and MnO content (<0.1%), and the highest Mg#=93 of the rock suite, with two groups of distinct Na concentration (Table 6; Fig. 5a, 5b). Diopsides from cryptic harzburgite are identical to the normal Na-type (Na_2O=0.25-0.45%) but with slightly higher Ti content (TiO_2 up to 0.2%; Fig. 5b). Diopsides from clinopyroxenites are also primitive in composition and closer to those of the harzburgite, but with slightly higher Ti content (TiO_2=0.2-0.4%). Plagioclase-dunites have two types of high Mg# clinopyroxene, with different minor elements compositions (Fig. 5a). One clinopyroxene type from plagioclase-dunite (sample 83DS-M11) has a composition (Mg#=91,

Table 7. Microprobe analyses of plagioclase in mafic rocks from the Terevaka transform fault and Pito Deep. Rock types include PLDU: plagioclase dunite, G: gabbro, LTR: leucotroctolite, MYG: mylonitic gabbro, MG: metagabbro, DIA: diabase, MIGO: olivine microgabbro, FG: ferrogabbro, v: vein in peridotite, z: wall rock reaction zone adjacent to contact. avg=average number of microprobe analyses, An=100*Ca/(Ca+Na).

Sample	*11DS*	*18DS-05*	*26*	*32*	*34*	*42*	*M11*	*M12*	*M14*	*M21*	*M28*	*M33*
Rock	*LTR*	*G*	*MG*	*G*	*DIA*	*MIGOv*	*PLDU*	*FGvz*	*MYGv*	*FGvz*	*FGv*	*Gz*
avg	*4*	*11*	*14*	*5*	*4*	*10*	*6*	*14*	*3*	*8*	*2*	*4*
SiO_2	46.36	51.50	51.36	52.04	54.65	53.98	47.83	62.12	53.88	61.07	60.01	56.34
Al_2O_3	33.57	30.48	30.21	30.31	26.77	29.24	33.03	23.86	29.71	24.51	24.82	28.33
FeO	0.36	0.31	0.18	0.20	0.74	0.11	0.15	0.21	0.04	0.26	0.13	0.11
MnO	0.02	0.02	0.03	0.03	0.02	0.03	0.01	0.07	0.02	0.03	0.02	0.04
MgO	0.15	0.02	0.02	0.02	0.12	0.02	0.05	0.01	0.02	0.06	0.00	0.05
CaO	17.45	13.44	13.55	13.77	10.89	12.09	17.09	5.70	12.34	6.24	7.18	10.81
Na_2O	1.64	3.88	3.88	3.93	5.06	4.82	2.02	8.56	4.85	8.01	7.79	5.52
K_2O	0.01	0.01	0.00	0.02	0.03	0.02	0.01	0.13	0.09	0.11	0.06	0.12
Total	99.60	99.72	99.29	100.40	98.37	100.37	100.22	100.73	101.04	100.34	100.13	101.42
O=32												
Si	8.578	9.393	9.408	9.431	10.042	9.730	8.768	10.967	9.662	10.832	10.700	10.009
Al	7.320	6.553	6.523	6.474	5.797	6.214	7.135	4.965	6.280	5.123	5.217	5.932
Fe	0.056	0.047	0.028	0.030	0.114	0.016	0.023	0.030	0.006	0.038	0.020	0.017
Mn	0.003	0.002	0.005	0.005	0.003	0.005	0.002	0.010	0.003	0.004	0.003	0.006
Mg	0.042	0.006	0.005	0.006	0.032	0.006	0.014	0.003	0.006	0.017	0.000	0.013
Ca	3.459	2.627	2.659	2.674	2.143	2.336	3.356	1.079	2.371	1.186	1.372	2.058
Na	0.588	1.373	1.379	1.380	1.802	1.686	0.717	2.930	1.687	2.753	2.694	1.902
K	0.003	0.003	0.002	0.005	0.008	0.004	0.003	0.028	0.020	0.025	0.014	0.028
An	85.41	65.62	65.82	65.88	54.20	58.04	82.33	26.72	58.15	29.93	33.64	51.60

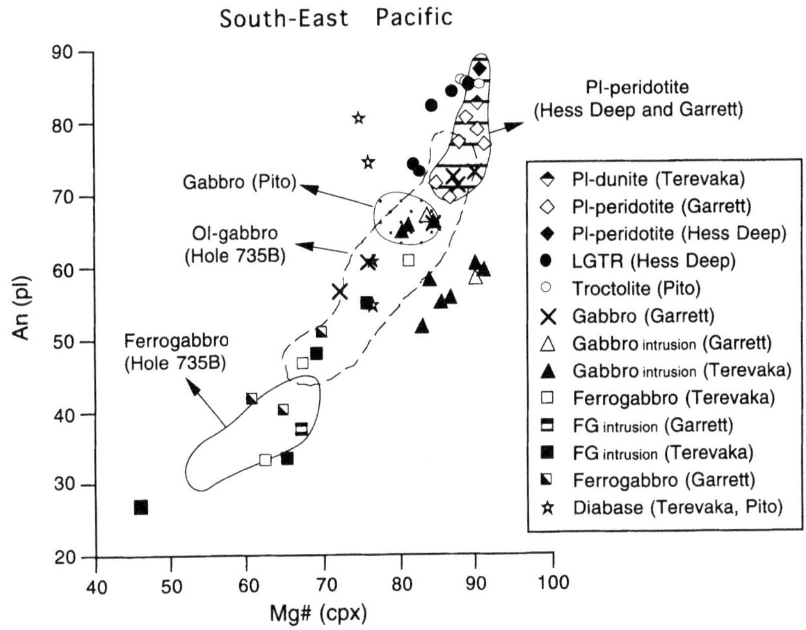

Figure 6. Covariation diagram of An content in plagioclase versus Mg# in clinopyroxene, from plutonic rocks of the South-East Pacific. Field with horizontal line defines plagioclase dunite from Hess Deep (Hékinian et al., 1993) and the Garrett transform fault (Hébert et al., 1983; Hékinian, Hébert, and Bideau, unpublished data). Fields for Hole 735B gabbroic rocks (AII fracture zone, South-west Indian ridge) taken from Hébert et al. (1991) and Constantin (1992). Plotted analyses represent averages per sample. All intrusions are in peridotite. LGTR= leucogabbro troctolite, FG= ferrogabbro. Gabbro intrusions in peridotite plot to the right of the main trend because their relatively high Mg# is due to chemical interaction with host ultramafic minerals.

Cr_2O_3=1.15% and TiO_2=0.37%) which is closer to the clinopyroxenite, whilst the other (sample 83DS-M18) is more diopsidic with Mg#=91.5, Cr_2O_3=0.6% and TiO_2=0.93% and is akin to the composition of clinopyroxenes from peridotite wall rock rims. Gabbroic veins in peridotite and massive gabbros from Terevaka have comparable clinopyroxene compositions except for higher Mg# in small-size veins (sample 83DS-28 and -M14 in Table 6) presumably caused by subsolidus Fe-Mg reequilibration (Figs. 5a, 6). Leucotroctolites from Pito Deep have clinopyroxene with higher Mg#, Cr_2O_3, Al_2O_3 and TiO_2 values than those from the gabbros (Figs. 5a, 6).

Plagioclase

Plagioclase spans a wide compositional range throughout the collection studied (Table 7, Fig. 6). Its composition varies from An30 in the intrusive ferrogabbro, via An50-65 in the intrusive olivine-gabbro, up to An75-An82 in the plagioclase dunite and leucotroctolite. The mineral composition of

Table 8. Microprobe analyses of amphibole in mafic rocks from the Terevaka transform fault and the Pito Deep. Rock types include PLDU: plagioclase dunite, G: gabbro, GTR: troctolitic gabbro, MG: meta-gabbro, MIGO: olivine microgabbro, FG: ferrogabbro, v: vein in peridotite, g: granoblast, z: wall rock reaction zone adjacent to contact. Mg#=100*Mg/(Mg+Fe).

Sample	*18DS-04*	*18DS-12*	*24*	*26*	*32*	*42*	*M12*	*M18*	*M28*	*M33*
Rock	*MG*	*GTR*	*Gz*	*MG*	*G*	*MIGOv*	*FGv*	*PLDU*	*FGv*	*Gz*
SiO_2	43.01	44.25	43.77	43.91	51.27	42.94	50.97	47.89	44.27	42.33
TiO_2	3.63	4.12	4.58	5.15	0.87	4.18	0.27	3.56	3.09	4.69
Al_2O_3	11.52	10.50	10.94	11.58	6.11	11.35	2.03	9.44	11.22	12.25
Cr_2O_3	0.59	0.40	0.98	0.80	0.30	1.21	0.00	0.82	0.14	1.24
FeO	10.60	9.29	4.87	5.33	8.56	7.43	23.76	4.63	10.17	8.64
MnO	0.07	0.04	0.10	0.11	0.09	0.08	0.64	0.14	0.31	0.13
MgO	14.50	15.41	17.68	16.86	18.70	15.89	11.29	18.62	15.00	14.60
CaO	11.87	11.07	11.02	11.94	12.34	11.77	9.91	11.58	10.79	11.52
Na_2O	2.50	2.86	2.88	2.36	1.52	3.02	0.77	2.74	2.90	2.92
K_2O	0.21	0.11	0.43	0.06	0.06	0.07	0.04	0.30	0.15	0.17
Total	98.50	98.05	97.23	98.14	99.82	97.94	99.67	99.73	98.03	98.49
O=23										
Si	6.247	6.393	6.280	6.241	7.153	6.203	7.585	6.643	6.413	6.115
Ti	0.397	0.447	0.494	0.550	0.091	0.454	0.030	0.371	0.337	0.510
Al	1.972	1.787	1.850	1.940	1.005	1.932	0.355	1.544	1.916	2.086
Cr	0.068	0.045	0.111	0.090	0.033	0.138	0.000	0.090	0.016	0.141
Fe	1.287	1.122	0.584	0.633	0.999	0.897	2.957	0.538	1.232	1.043
Mn	0.008	0.005	0.012	0.013	0.011	0.009	0.081	0.017	0.038	0.015
Mg	3.138	3.318	3.781	3.572	3.888	3.421	2.505	3.849	3.238	3.144
Ca	1.847	1.714	1.694	1.818	1.845	1.821	1.580	1.721	1.674	1.783
Na	0.705	0.802	0.802	0.651	0.410	0.845	0.222	0.737	0.814	0.818
K	0.039	0.021	0.079	0.011	0.010	0.013	0.007	0.053	0.027	0.031
Mg#	70.92	74.73	86.62	84.95	79.56	79.23	45.86	87.74	72.44	75.09

the leucotroctolite from Pito Deep is primitive (An85, Fo86, diopside Mg#=89), and equivalent to those from Hess Deep and the Garrett transform fault (Hékinian et al., 1993, and unpublished data).

Amphibole

Kaersutite and pargasite (Table 8) are most often found disseminated interstitially amidst the plagioclase + clinopyroxene matrix in intrusive microgabbros (83DS-42), or as discrete foliated bands in ferrogabbro intrusions (83DS-M28). Kaersutite inclusions do also occur occasionally in chromite of plagioclase dunite (83DS-M18) and in gabbro/peridotite reactive layers (83DS-24). In the gabbros, kaersutite generally occurs interstitially around clinopyroxene (18DS-12).

Harzburgite

Plagioclase-free peridotites from the Terevaka transform fault are harzburgites which generally have a porphyroclastic to mylonitic texture (Plate 1G). Olivine and orthopyroxene are 1-10 mm long and show high-temperature deformation features such as kink-bands, deformed cleavages and recrystallized grains. The primary minerals of the harzburgites have a strongly residual composition: olivine (Fo=91, NiO=0.38-0.40%), enstatite (Mg#=90.8, Al_2O_3=2.5%, Cr_2O_3=0.72%, TiO_2<0.03%), spinel (Cr#=40-50, Mg#=70-60, TiO_2<0.1%), and trace amount of very small (50-100 mm) disseminated granular diopside. Two groups of diopside are recognized based on Na content (average Na_2O=0.06% and 0.38%; Fig. 5b) for otherwise similar element contents characteristic of a refractory composition (Mg#=93.1, Wo46.4, Al_2O_3=2.8%, Cr_2O_3=1%, TiO_2<0.08%). In one case, the two diopsides coexist in the same sample (83DS-28; Table 6). Other variations exist among the samples, with small spinel granoblasts showing lower Cr#=38 at comparable Mg#=64 for larger spinel grains of the same sample (83DS-43; Table 3). A subset, called cryptic harzburgite, designates the analyzed zones located within 1 cm of any intrusion but away from the immediate wall rock contact. In the cryptic harzburgite, notable modification of the host rock is recorded in the chemistry of spinel (Fig. 2) and diopside (Fig. 5).

In general, the mineral compositions are close to those of the harzburgites from the Garrett transform fault (Hébert et al., 1983; Hébert et al., 1989; Cannat et al., 1990; Hékinian, Hébert, Bideau and Constantin, unpublished results) but, for similar Mg# ratios, spinels have lower Cr# ratios in comparison with peridotites from Hess Deep (Cr#=50-60; Hékinian et al., 1993; Girardeau and Francheteau, 1993).

Clinopyroxenite

Three samples show massive clinopyroxenitic veins, and clusters made of small granular minerals (>90% diopside and minor enstatite) set within harzburgite (Table 2). In those samples, diopside-rich parts are well preserved from the serpentinisation seen in the harzburgitic parts. In addition, well-preserved specimen of large enstatite crystals along lithological contacts provide indications of replacement textures. One of these samples (83DS-M163) is a 5 mm thick clinopyroxenite vein oriented sub-parallel to the foliation in the host peridotite (Plate 1A). Compared to the host harzburgite, diopside from this clinopyroxenite has on average slightly lower Mg#=92.4, Al_2O_3=2.51% and higher TiO_2=0.37% (Fig. 5), for otherwise similar element contents (Table 6). Small foliated anhedral diopside grains (100-200 mm) display a granoblastic, mosaic texture with minor amounts of very small intergrowths of enstatite and spinel. The second sample contains a 2 cm cluster of clinopyroxenite with a similar fine-grained granular texture occurring in the harzburgitic part of an intruded and impregnated peridotite (Plates 2C, 2D, 3H; sample 83DS-M28). This cluster is made of over 90% of anhedral porphyroclasts of diopside with high Mg#=92.8%, Al_2O_3=3.94%, Cr_2O_3=1.07%, TiO_2=0.25% and Na_2O=0.39%, with an interstitial, very fine-grained assemblage of orthopyroxene + spinel ± olivine. The contacts are parallel but wavy, and are marked by replacement of the harzburgite mineralogy where orthopyroxene porphyroclasts from the wall rock are partly replaced by the small granular diopsides (Plate 3H). The third clinopyroxenite-bearing sample (Plates 2F, 3E, 3F; sample 83DS-M33) shows a gradual modal and compositional variation which illustrates a sequence of metasomatic events. Along a 3 cm long traverse starting symmetrically from both borders, the specimen is made up of medium to coarse grained, plastically deformed Cr- and Al-rich diopside (Wo45.8, Mg#=91, Al_2O_3=3.9%, Cr_2O_3=1.3%, TiO_2=0.20%, Na_2O=0.42%) coexisting with holly-leaf spinel (Mg#=61.6, Cr#46.8, TiO_2=0.28%) and stretched and partly disaggregated olivine clusters (Fo88.8, MnO=0.18%). Diopside, in turn, is partly corroded and replaced by abundant small oriented, tablet-shaped augite granoblasts (Wo43.4, Mg#=87.9, Al_2O_3=3.66%, Cr_2O_3=1.08%, TiO_2=0.20%, Na_2O= 0.41%; Plate 3F). Within the middle part of the sample, the mineral assemblage is gabbroic, with coexisting augite + plagioclase + kaersutite + interstitial sulfides (mainly pyrrhotite and pentlandite). From the middle gabbroic part to the olivine-bearing clinopyroxenite there is a strong gradient of wall-rock reaction, suggesting that there has been at least one stage of porous-flow metasomatism caused by circulation of a fluid-bearing melt.

Plagioclase dunite

The plagioclase-dunite is characterized by an olivine-rich granular matrix with minor orthopyroxene, and interstitial oriented plagioclase with associated clinopyroxene (sample 83DS-M11; Plate 1C). The plagioclase forms lenses, a few cm in length, oriented oblique to the olivine + orthopyroxene foliation. Spinel is black, included in plagioclase, and shows evidence of disequilibrium with irregular contours (Plate 3A). Another type of plagioclase dunite (sample 83DS-M18) is marked by irregular plagioclase-rich veins (Plate 1E) which have pervasively penetrated the peridotite matrix. In these two samples, the plagioclase foliation makes high angles with the olivine + orthopyroxene foliation. A third type, not shown here, is made of heterogeneously distributed zones of plagioclase-bearing and plagioclase-free dunite.

Plagioclase dunites from Terevaka have mineral compositions intermediate between those of plagioclase-free peridotite and the primitive leucotroctolite from the Pito Deep (Figs. 3, 5 and 6). The mineral chemistry reveals their composite nature, with orthopyroxene and olivine being homogeneous, while plagioclase and clinopyroxene from the same sample have heterogeneous compositions, in particular in sample 83DS-M18 (Table 2). Augite has a more heterogeneous Ti-rich, Cr-poor composition (Fig. 5; TiO_2=0.69%, Al_2O_3=2.67%, Cr_2O_3=0.64%, Mg#=91.7) than those from the matrix (TiO_2=0.41%, Al_2O_3=3.49%, Cr_2O_3=1.29%, Mg#=90.5). This may be explained by the fact that minerals inside veins (plagioclase and clinopyroxene) were the last to crystallize. Our observations suggest that most of the plagioclase and clinopyroxene crystallized from percolating primitive melt flowing through the peridotite, and that the plagioclase and clinopyroxene in the veins crystallized later from trapped and/or transient, evolved melt compositions.

Gabbroic intrusions and wall-rock reactions in peridotite

In this section we compare the nature and composition of the various gabbroic intrusions with the massive gabbroic rocks from the Terevaka transform fault and, also, from other known oceanic localities. Particular attention is paid to harzburgite wall rock adjacent to intrusions, where chemical disequilibrium led to mineralogical and chemical transformations.

Intrusive gabbros and ferrogabbros

On the basis of their order of appearance, and coupled with the mineral chemistry, the intrusive gabbros of the Terevaka transform fault are separated into two main groups, i.e., the primitive oxide-free gabbro group (including olivine-microgabbro, gabbro, and mylonitic gabbro) and the late ferrogabbro and mylonitic ferrogabbro group. The structural and textural characteristics of the gabbroic vein types from the Terevaka transform fault and their principal mineral constituents and compositions are listed in Table 2. The distinction between oxide-free and oxide-bearing gabbros is important, since these two groups crystallized from respectively basaltic and ferrobasaltic parental magmas leading to different metasomatic effects on the host peridotite. Notably, the spinel compositional trend extending toward low Mg# (as low as 20), high Cr# (up to 68, Fig. 2a) and very Ti-rich (TiO_2 up to 6%, Fig. 2b) concerns grains along the wall rock margins of ferrogabbroic veins but still within the host peridotite (Fig. 2). This Fe-Ti-enrichment trend is limited by the disappearance of spinel as a stable phase in the adjacent ferrogabbro, where abundant Fe-Ti oxide occurs. When compared to other abyssal localities (Dick and Bullen, 1984; Hébert et al., 1989), these spinels are by far the richest in TiO_2 found in the ocean basins, and reflect the extreme chemical disequilibrium between the fractionated ferrobasaltic melt and the host peridotite (Fig. 2b).

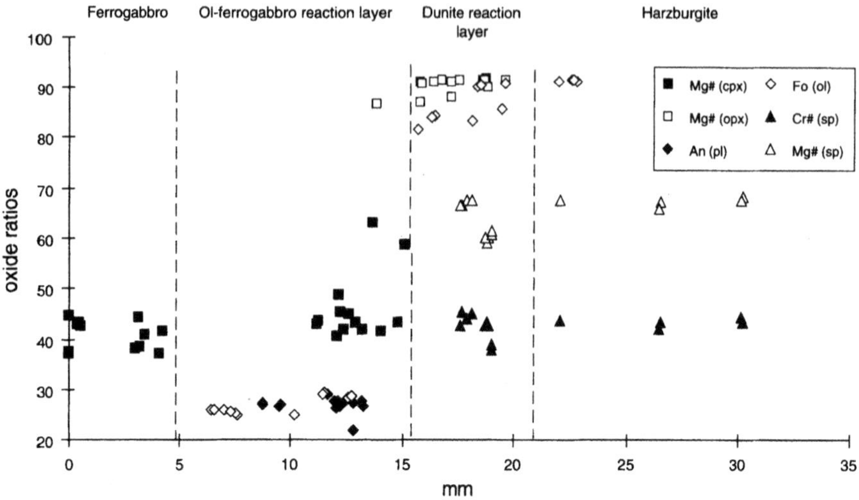

Figure 7. Detailed microprobe traverse across ferrogabbro/peridotite contact zone in sample 83DS-M12 (Plate 1B). Dashed lines indicate boundaries of reaction zones developed between host harzburgite (right-hand side) and porphyroclastic apatite-bearing ferrogabbro intrusion (left-hand side). Two types of reaction layers are developed in between these two lithologies (Plate 1B): a 1.7 cm-thick black dunitic layer (Fo84), and a 0.6 cm thick olivine- and ilmenite-rich ferrogabbro layer (Fo27, An27; see Plate 3C). No olivine has been found in the ferrogabbro. Note that spinel is unstable and replaced by Fe-Ti oxides. In the dunitic layer, two types of orthopyroxene are found: large enstatite porphyroclasts with Mg#=90-91, corroded by small- to medium-grained orthopyroxene (and olivine) granoblasts showing progressively decreasing Mg# (90-85) towards the olivine-ferrogabbro layer. A similar trend has been noted in the Al_2O_3 contents of orthopyroxene granoblasts which decrease from 2.8% to 1.7% (Fig. 4d), presumably because of competition by crystallizing plagioclase in the adjacent olivine-ferrogabbro layer.

Most of the gabbros show an unequivocally intrusive relationship with the peridotite (e.g., Plate 2A) but this relationship is not immediately obvious in all cases. For example, the intrusive character of some ferrogabbro pockets sub-parallel to the foliation in the peridotite is inferred from the presence of sub-millimeter scale magmatic veinlets in the peridotite which root in the gabbro, the presence of a reaction layer along the wall rock contact zone, and the presence of detached and partly corroded dunitic clasts within the ferrogabbro (e.g. sample 83DS-M21, Plate 2B). This last observation implies that some intrusion resulted from magmatic injection and crystallization accompanied by solid mineral flow and extensive subsolidus recrystallization. In fact, clinopyroxene from the most deformed ferrogabbros have very low Al^{IV} (<0.03) and Ti (<0.01) contents (Table 6) which are affiliated with metamorphic compositions (Mével, 1987; Hébert et al., 1991).

When compared to massive gabbros from the same locality, the totality of the intrusive gabbros show a wider range of lithological variations (from oxide-free to oxide-rich gabbros) and mineral compositions. In particular, plagioclase and clinopyroxene of the intrusive gabbros have a large compositional range which cover both extremities of the oceanic spectrum (Fig. 6).

It follows that the gabbroic intrusives from Terevaka have crystallized from liquids ranging from primitive to fractionated compositions. Ferrogabbros have heterogeneous An and especially Mg#, which may be explained by the infiltration of TiO_2- and FeO-rich melt percolating through and reequilibrating with pre-existing gabbro and peridotite. It is remarkable that the intrusive gabbros from this single locality (Terevaka) also cover the entire compositional range of the 500 m deep Hole 735B gabbroic rocks (Fig. 6). This feature demonstrates the importance of processes leading to the isolation of parts of the magmatic lense from its supply zone and allowing it to fractionate regardless of the host rock (peridotitic at Terevaka, gabbroic at Hole 735B). Moreover, an interesting parallel is made between the compositions of olivine reaction layers at ferrogabbro-peridotite contacts sampled from the Terevaka fracture zone (Fig. 7; sample 83DS-M12) and the most evolved olivine compositions in ferrogabbros from Hole 735B (Hébert et al., 1991; Constantin, 1992; Constantin and Hébert, 1992). Both olivines have approximately Fo30, polygonal grain shapes and granoblastic textures, suggesting that they are the products of reactions. In addition, the ferrogabbros in both cases are apatite- and zircon-bearing, indicating a highly differentiated nature.

Interstitial, brown Ti-bearing amphiboles (kaersutite and pargasite in Table 8) in various gabbros intrusive into peridotite are inferred to be of late-magmatic origin and to have developed during the last stages of crystallization when the magma had fractionated and contained a higher amount of volatiles (Hébert and Constantin, 1991). Experimental determination of pargasite stability relationships in the presence of orthopyroxene supports the hypothesis that brown amphibole could form from a hydrous silicate melt at depths as shallow as 8-10 km within the oceanic upper mantle (Lykins and Jenkins, 1992). This suggests the circulation, during the last magmatic episode, of late-magmatic hydrous fluids deep into the oceanic lithosphere.

Peridotite wall-rock reaction and dunite veinlet

The range of compositions of the intrusive gabbros must be controlled by the variability of parental magmas and subsequent crystal fractionation events occurring at various melt/rock ratios. The mineralogical transformations seen along intrusion fronts (Plates 1B, 2B, 2E, 3C) and in zones between several veins (black dunitic part of plate 1F), presumably as a result of extensive impregnation (black dunitic mass in Plates 1C, 1E), clearly indicate reactions between intruding melts and host rock. All magmatic vein types listed in Table 2 are associated with transformations in harzburgite, mainly in the form of granoblastic, fine-grained peridotite clusters or rims, suggesting wall-rock reactions involving orthopyroxene dissolution and olivine crystallization. For example, evidence of reaction is seen in several samples (83DS-M12, -M21) of evolved ferrogabbro (olivine with Fo27) in contact with a harzburgite (plates 1B, 2B, Fig. 7). The two rock types are separated by two different reaction zones, one gabbroic and the other dunitic (Table 4). The dunitic harzburgite shows fine grained equigranular olivine-rich clusters and/or medium grained granoblastic rims near the gabbroic intrusion which are implied to be reaction products (Plate 3C). Peridotite reaction zones have granoblastic textures and are made up of small polygonal olivine grains (Fo=83.7, NiO=0.34%) showing triple junctions of neighbouring grains. These olivines replace large kinked orthopyroxene grains with mantle compositions (Mg#=91, Al_2O_3=2.5%, Cr_2O_3=0.69%) along irregular edges. Spinel from the peridotite reaction zone has a different composition (Mg#=60, Cr_2O_3=35%, Cr#=43.3) when compared to host rock spinel which has a typical mantle composition (Mg#=67.2, Cr_2O_3=37.8%, Cr#=43.3). In addition to these reaction rims, dunite veinlets of granoblastic olivine with distinct compositions (Fo90-91, NiO=0.24%; Fig. 3A) crosscut host peridotites near contacts with intrusive gabbro (Plates 1F, 3B). In view of the immobility of Ni in olivine, this composition is explained by crystallization of olivine from an infiltrating basaltic melt with Fe and Mg contents buffered to high Fo values by minerals in the harzburgite. This effect is probably accentuated with subsolidus Fe and Mg exchange.

In summary, peridotites from the Terevaka transform intruded by ferrogabbroic veins show general enrichment in Ti and Fe (Fig. 7). In particular, the compositions of spinel from wall rock peridotite are modified from those of spinels located away from veins but also in harzburgite, by removal of Al and Mg and addition of Fe and Ti (Fig. 2). Moreover, clinopyroxene has higher Ti contents and lower Mg# than diopside in harzburgites (Fig. 5). Magmatic dunitic reaction rims are found along contacts between mantle peridotites and evolved ferrogabbroic intrusives (Plate 2B). Sub-solidus reequilibration processes through Fe-Mg exchange have locally affected the compositions of coexisting minerals in the impregnated peridotites, lowering the Mg# content of the olivine, orthopyroxene and spinel whilst the Mg# of clinopyroxene in gabbroic veins has raised (Fig.7). This is probably due to buffering by the host peridotite on the minerals crystallizing in the veins, in particular in the case of small-scale veinlets. For both gabbroic intrusive types, the resulting effect is to produce a general dunitic enrichment in the host peridotite by the reaction: enstatite + basaltic melt → olivine.

The above small-scale magmatic processes resulting in olivine enrichment bear similarities to those originally described by Boudier and Nicolas (1972) and Quick (1981) to explain large dunitic pods in ophiolite complexes, and by Berger and Vannier (1984) on the origin of dunitic enclaves in oceanic island basalts. The occurrence of several intermediate mineral compositions in the present case (Table 4) probably reflects the diversity of intruding melts and the extreme compositional differences between melt and host rock.

Discussion

Melting and deep refertilisation processes at ocean ridges

Experimental studies on mineral stability and composition in the upper mantle indicate that, during melting, compatible elements such as Cr, Ni and Mg remain in the residue while more incompatible ones such as Ti, Mn, Na, K and Al migrate into the melt (Hirose and Kushiro, 1993; Takahashi et al., 1993). For high degrees of fusion, modal clinopyroxene contents of residual peridotite decrease while olivine contents increase. Mineral compositions of abyssal, plagioclase-free harzburgite have been used as indicators of melting processes under spinel- to plagioclase-lherzolite stability facies. When compared with the compositional field of abyssal peridotites from slow spreading ridges presented by Dick and Bullen (1984), the spinels from the Terevaka transform fault plot in the high Cr# ratio field which is inferred to result from large degrees of fusion (Fig. 2a). Moreover, spinel peridotites from Terevaka have a residual harzburgitic character with very low modal clinopyroxene contents. In contrast, peridotites along the MAR have relatively high clinopyroxene contents and lherzolites are common (Bonatti et al., 1992; Shibata and Thompson, 1986, Michael and Bonatti, 1985). The spinel compositions of Terevaka transform fault peridotites and of other fast spreading centers along the EPR in the Garrett transform fault and from Hess Deep are used to decipher regional mantle trends. The Pacific spinel compositions suggest that the degree of fusion is relatively high, on average higher than at slow spreading ridges where the Cr#=10-55 ranges to lower values (Dick and Bullen, 1984).

The chemical variations among the sampled plagioclase-free harzburgites are clearly reflected in the pertinent spinel and diopside compositions (Tables 3, 6; samples 83DS-28, -43). The two groups of diopside showing different Na contents (average Na_2O=0.06% and 0.38%, respectively; Fig. 5b) at otherwise element contents characteristic for clinopyroxenes of strongly residual origin (Mg#=93.1, Wo46.4, Al_2O_3=2.8%, Cr_2O_3=1%, TiO_2<0.08%) are interpreted as fossilized witnesses of the superposition of post-melting processes. In one case, the two diopsides coexist in one and the same sample 83DS-28 (Table 6), which probably reflects variable but small amounts of trapped and/or transient melts. As noted by Kornprobst et al. (1981), clinopyroxene is the only Na-bearing phase in the spinel-peridotite facies, hence its composition should reflect the Na content of the whole rock. The present distinction of two groups of diopside with different Na contents has important implications for the interpretation of melting processes, since Na is used as a petrological indicator in both MORB (Klein and Langmuir, 1987) and peridotites (Elthon, 1992). In the sample collection studied, diopside with the higher Na_2O content is the most common, and its Na value corresponds to the average value of oceanic harzburgite (Dick, 1989). Elthon (1992) reinterpreted the peridotite data of Johnson et al. (1990) as being not pure residues of fusion but residues refertilized by basaltic melt due to some kind of cryptic metasomatism. It is, therefore, possible that most of the diopsides in oceanic harzburgites analyzed so far are the product of various amounts of refertilisation due to cryptic metasomatism, and that only the low-Na type of this study (Fig. 5b) represents a pure mineral residue of melting. In any case, the fact that the unimpregnated harzburgites from Terevaka have very low modal diopside indicates that they were the subject of an extensive melting event.

Origin of the intrusions in harzburgite

In our following discussion on the origin of the Terevaka transform fault intrusions, we feel that the numerous field studies and pertinent models concerned with lower crustal and upper mantle sections of ophiolites provide an essential basis. For example, field, structural and petrological studies of the crust-mantle transition zone in a number of ophiolite massifs (Quick, 1981; Nicolas and Prinzhofer, 1983; Evans, 1985; Reuber et al., 1985; Nicolas, 1986; 1989; Bédard 1992; Ceuleneer and Rabinowicz, 1992) have documented several types of interactions between basaltic melt and residual peridotite, and these interactions have principally been related to the HOT (harzburgitic) type ophiolite massifs as defined by Boudier and Nicolas (1985) and Nicolas (1989). The stratigraphic succession of intrusive rock types, described by Nicolas (1986, 1989) and his coworkers from the Oman ophiolite, varies over distances of a few to several hundreds of meters, and grades from a dunite and orthopyroxenite-rich sequence via websterite-clinopyroxenite and clinopyroxenite-gabbroic rocks to an ultramafic-mafic transition zone of variable thickness interpreted as the crust-mantle boundary. The authors have shown that the rheology of the host harzburgite did progressively change from viscous to brittle, thereby controlling the injecting melt and its chemical evolution. Nicolas (1986, 1989) has noted that basaltic melts could circulate through peridotite in two ways, i.e., (1) by percolation along grain boundaries (porous flow) and (2) by brittle fracturing and injection through veins and dikelets. Consequently, Nicolas (op. cit.) and coworkers have classified dikes as "in situ" or "intrusive", the first type being intruded into a melt-bearing peridotite (around 1200 °C), whilst the second type is emplaced at much lower temperatures around 600 °C at which the host peridotite is essentially brittle.

By analogy with these observations in HOT ophiolites, we infer a sequence of magmatic injections in the harzburgites from the Terevaka transform fault (Table 2) which records a similar evolution. We thus infer that magmatic injections started relatively deep (> 8 kbar) with the formation of clinopyroxenite veins under high-temperature conditions of the spinel-lherzolite facies, and that these injections were followed, during exhumation into the plagioclase-lherzolite facies, by the formation of plagioclase-dunite and the intrusion of olivine gabbros. Eventually, the cooling environment of the peridotite promoted fractionation of MORB, and brittle conditions became favorable for the intrusion of ferrogabbroic compositions.

Origin of the clinopyroxenite

The clinopyroxenite vein from Terevaka has a major element content in equilibrium with mantle compositions (Table 6, sample 83DS-M163). The clinopyroxenite is a highly deformed, porphyroclastic to locally very-fine grained mylonitic rock (Plate 1A). The precise origin of the clinopyroxenite is difficult to assess and lies beyond the scope of the present paper, but we wish to mention three hypotheses, i.e., high-pressure fractionation of primitive melt flowing through partially melting harzburgite, suction into the vein of in-situ liquid from the melting peridotite, and deep-seated metasomatic replacement of harzburgite by melt infiltration. The refractory composition of its mineralogy and the absence of plagioclase as a stable phase suggest that the clinopyroxenite from Terevaka formed in the spinel-lherzolite stability field (> 8 kbar). The textural features suggest that the harzburgite is metasomatically replaced by a melt reacting with the host mineralogy (Plates 1A, 2C, 2D, 3H). The clinopyroxenite vein in an almost diopside-free harzburgite may thus represent a fossil example of melt aggregation and reaction in the partially molten upper mantle, at depths larger than 25 km (> 8 kbar). Clinopyroxenites are seldomly sampled in the ocean basins due to the inherent difficulty to access the pertinent mantle sections and because of the general overprinting by serpentinisation. Nevertheless, clinopyroxenites and diopside-rich peridotites are important clues to decipher deep melt extraction processes and mantle heterogeneities. We hope that work in progress using an ion probe to determine the trace element chemistry in the diopside will help to constrain the process responsible for these puzzling clinopyroxene-rich compositions.

Origin of dunitic and associated mafic rocks

Between 40 and 60% of the peridotite samples recovered in the dredge haul (83DS) show evidence of basaltic interaction and impregnation, forming veins and pods of gabbros and trace pyroxenite. This locally very important phenomenon was also noted by Dick (1989). Following field studies in ophiolites by Nicolas (1986), Dick (1989) has interpreted the abyssal plagioclase-bearing peridotites as the product of impregnation of depleted, residual peridotite by either in-situ or transient melt. With the assumption that the normative plagioclase content (55-60%) of abyssal basalts is uniform, Dick (1989) used the plagioclase content of abyssal peridotite to estimate the minimum amount of trapped melt, and reproduced the major element composition of plagioclase-bearing peridotites by simple mixing of basaltic melt and residual peridotite.

Foliated and mottled plagioclase dunites from the Terevaka transform fault also result from the percolation and impregnation of variably fractionated basaltic liquid in peridotite (Constantin and Hékinian, 1993). The plagioclase-peridotites from the Terevaka transform fault (83DS-M11, -M18) have major element bulk compositions (not shown) which can be reproduced by adding 24% of a basaltic composition (diabase 83DS-38) to 76% harzburgite (peridotite 83DS-20). Further support for this interpretation of the origin of abyssal plagioclase peridotites comes from experimental studies on the interaction between harzburgite and basaltic melt which produces a plagioclase-bearing dunitic product (Boudier, 1991). Spinels in plagioclase-dunites from Terevaka have unusually high Cr# (Fig. 2a) caused by preferential incorporation of Al in plagioclase. From experimental data, Roeder and Reynolds (1991) have related this large change in spinel-Cr# from MORB to the Al_2O_3 content of the melt which, in turn, is highly sensitive to the crystallization or melting of plagioclase. They also noticed that the Ti content of chromite increases with Fe enrichment and decreases with temperature. In spinels from intruded peridotite wall rock we observe a strong relationship between their compositions and the gabbroic versus ferrogabbroic nature of the intrusive (Fig. 2a, b). Previous studies of spinel compositions in plagioclase-bearing ultramafic rocks have shown similar evolutionary trends with increasing TiO_2 and Cr#, and decreasing Mg# of the peridotite reacting with basaltic melt (Henderson, 1975; Ridley, 1977; Berger and Vannier, 1984). Other processes of formation of plagioclase-bearing peridotites involve the sub-solidus metamorphic reaction of spinel due to decompression (Hamlyn and Bonatti, 1980; Rampone et al, 1993). This closed system process induces characteristic exsolution and coronitic textures which are not found in our samples. The coalescent aspect of plagioclase lenses in our plagioclase-dunites (Plates 1C, 1E; samples 83DS-M11 and -M18), and the dissolution textures suggested by corroded anhedral black spinel (Plate 3A) point to melt/crystal disequilibrium. We infer that reaction crystallization (hybridization) between basaltic melt and residual peridotite is probably closer to the real conditions of formation of plagioclase-peridotite. Kelemen (1990) and Kelemen et al. (1990, 1992) postulated that dunitic reaction zones form by interaction between solid peridotite and silicate melt infiltrating along grain boundaries. They noticed that olivine produced by melt/peridotite reactions falls in the restricted compositional range of Fo83-89, i.e., the same range as we observe in our plagioclase-bearing ultramafic rocks (Fig. 3b). As mentioned previously, dunites from the Terevaka transform fault are always associated with gabbroic intrusions and impregnations. These features provide strong evidence that dunites are the product of melt/peridotite interactions, and that they are not pure residues of large degrees of melting.

As to the leucotroctolites from the Pito Deep, their composition, texture, and in particular the poikilitic habit of the clinopyroxene also suggest that they represent shallow-level hybridized cumulate rocks formed in response to impregnation of a large volume of basaltic melt into residual peridotite. The resulting mush would then follow fractional crystallization, with expulsion of residual liquid such as postulated for crustal sections of the Oman ophiolite (Benn et al., 1988). This hypothesis may explain the discrepancies, found at high An, Mg#, and Fo (Fig. 6) mineral contents, between natural and experimentally crystallized mafic plutonic rocks (for a review, see Elthon et al., 1992) derived from normal tholeite. We consider this a plausible way of producing primitive oceanic plutonic rock without requiring unusually primitive parental compositions for the natural samples. Possible analogues of those plutonic rocks may occur in the North Arm massif of the Bay of Islands ophiolite, where a large part of the crustal section has an hybrid origin (Bédard and Constantin, 1991; Bédard, 1992).

Melt transport by intrusion

The physical properties of host rock and intrusion, and the nature of the intrusion-host interaction mainly depend on porosity and cohesion of the materials which, in turn, are at least partly controlled by ambient pressure and temperature. Like in ophiolite field studies, we observe that the injecting melt is either confined to produce gabbroic intrusions or that the melt is dispersed and percolates through the host peridotite. The sequence of intrusions emplaced in residual harzburgites from the Terevaka transform fault also corresponds with a gradually changing rheological behaviour. The fracturation-injection of basaltic liquid leading to vein networks in the host peridotite is a dynamic phenomenon. The crystal aggregates, found in intrusive gabbroic veins of this study, do not represent bulk melt compositions. Instead, they are solid residues of mixing and fractional crystallization which, comparable to cumulates, may episodically modify the melt composition as the melt flows through the veins. The mineral chemistry of the phases present along intrusive contacts of fractionated gabbros in peridotites (Fig. 2b, 7) clearly shows melt-crystal disequilibrium. The extreme compositional and modal variations, even at the scale of a thin section, presumably reflect the percolating nature of the process. Local bulk compositions of the host produces large chemical gradients, and influences the compositions of the crystallizing minerals (in particular Mg#; Fig. 2a, 3b, 6). It follows that these mineral compositions reflect time- and temperature-dependent processes which influence melt composition and melt/rock ratios. Percolating primitive melts could equilibrate with the invaded peridotite matrix, such that percolation (chromatographic column type) will also influence melt evolution (Harte et al., 1993). Nicolas (1986) has noted that dikes and veins at Moho levels are likely to remain semi-stagnant for periods of time, and that enhanced chemical exchange with the wall rock is likely to occur during times of stagnation (Wilshire and Kirby, 1989). Examples of vein networks show that veins may mutually crosscut, hence, that while some veins at a given locality are open, nearby others may close (Plates 2A, 2B, 2E). Such small-scale mechanical variations of closed and open system behavior will likely affect magma compositions. The scale and width of veining, the intensity of the deformation and the available surface area are clearly important factors in the ability of melts to percolate through and react with the peridotite. The gabbro veins of similar modal and mineral composition (Table 2) may either be undeformed, and crosscut the host peridotite foliation, or they have porphyroclastic to mylonitic textures and become parallel to the foliation in the host peridotite (Plates 1D, 1G). Sleep (1988) has noted that near the ridge axis deviatoric strain rates are low so that porous flow should dominate, and that away from the ridge axis increasing deviatoric strain rates should promote vein and dike growth. Likewise, both Ildefonse et al. (1993) and Quick and Delinger (1993) have suggested that feeder dikes that penetrate the cumulate pile near the spreading axis will be rotated parallel to the flow foliation, but those that invade farther from the ridge will be less deformed and retain crosscutting relationships.

Tectonic and magmatic implications of sea-floor exposure of intruded peridotites

Recent models for the emplacement of mantle rocks on the seafloor of slow-spreading ridges such as proposed by Cannat (1993) envisage a discontinuous magmatic crust, involving a magma and heat supply too low to produce a 4-7 km-thick crust. It is suggested that alternating magmatic and tectonic episodes would cause upper mantle material to be tectonically exhumed through the axial lithosphere, where it is intruded by short-lived gabbro lenses and pockets (Mével et al., 1991; Cannat, 1993). Extensive gabbroic intrusion in the Terevaka transform fault peridotites indicate a locally thin crust. Compilation of seismic refraction results by White et al. (1992) point to an average thickness of the oceanic crust of 7.1 ± 0.8 km, with extreme lower and upper bounds of 5 and 8.5 km. These estimates apply to normal oceanic crust away from anomalous regions such as fracture zones and hot spots. At deep transform faults, the crust is usually thinner, and mafic and ultramafic plutonic rocks are favorably exposed on the floor and walls of the valley. This situation could preferentially occur at ridge-transform intersections where the crust is inferred to be thin (Fox and Gallo, 1984; White et al., 1992).

The gabbroic intrusions in the Terevaka transform fault peridotites may have been formed by melts flowing laterally in an along-axis direction, as suggested by Natland (1980) for the Galapagos rift and also invoked for both the EPR and MAR segments (Francheteau and Ballard, 1983; Karson and Dick, 1984; Batiza and Niu, 1992; Sinton and Detrick, 1992). Intuitively, it seems reasonable to argue that melt flowing in an along-axis direction towards the fracture zone has a greater potential to intersect uplifted lithospheric mantle sections. In agreement with known petrological and structural effects of ridge-transform intersections (Langmuir and Bender, 1984; Fox and Gallo, 1984), the intersection with the fracture zone may represent the distal and cooler end of the ridge segment. Melt or mush in such a setting are thus located away from the main recharge center of a magmatic lense. This scenario is similar to models of oceanic magma chambers envisaged by Sinton and Detrick (1992), and is supported by recent thermal, mechanical and geophysical studies which integrate intrusive processes in the construction of the oceanic crust (Phipps Morgan and Chen, 1993; Henstock et al., 1993). Such magma, being isolated at depth from its supply reservoirs, evolves to a fractionated ferrobasaltic composition. The concurrent effect of cooling and increasing plastic deformation is expressed in the progressive development of ductile shear zones. We believe that, during the last magmatic episode, the ferrogabbroic mush intruded in successive injections as small sills, dikes and veins at a stage that the peridotite had cooled to subsolidus temperatures as suggested by its brittle behaviour. The highly fractionated composition of the ferrobasaltic liquid and its fluid content, inferred from the presence of primary brown amphibole, probably caused a solidus temperature lower than that of a more primitive gabbro. Local variations of melt/rock ratios and temperature may account for the joint occurrence, at a single locality, of abundant basaltic impregnation as well as dikes and veins. It is likely that the presence of liquid impregnations in the upper mantle and lower crust is related to a dynamic process occurring preferentially in amagmatic accreting ridge segments, and to episodically alternating magmatic and tectonic phases (Cannat, 1993; Bideau and Hékinian, 1994).

In search for complementary mechanisms other than purely tectonic uplift or diapiric serpentinization to explain the exposure of upper mantle rocks on the sea-floor, we propose as a tentative hypothesis that intrusion of gabbroic melts into upper mantle peridotites could play a triggering role in mantle exhumation. Intrusions occur early in the accretion process and possibly concurrently with tectonic uplift. Extensive basaltic intrusion will tend to lower the mean density of a given volume of upper mantle rocks affected by intrusion. Batiza and Niu (1992) calculated densities for EPR 9°30'N melts of around 2.7 g/cm^3, which is much lower than the mean oceanic mantle density of 3.3 g/cm^3 calculated by Solomon and Toomey (1992). Even a crystallized gabbroic melt has an average bulk density of 3 g/cm^3 as reported for ODP Hole 735B gabbros (Dick et al., 1991), which is still lower than the mantle average. We suggest that, by virtue of their low density and brittle behaviour, these gabbroic bodies inside the peridotite constitute preferential zones of weakness favoring localized faulting by which entire massifs are brought to the sea-floor.

Conclusions

This study shows how a homogeneous, residual oceanic harzburgite may be intruded and locally metasomatized by various amounts of melts and fluids, starting at high magmatic temperatures decreasing all the way down to low-temperature seawater alteration. By emphasizing observations at a single locality and concerning magmatic processes occurring at the microscale, we demonstrate the locally highly heterogeneous nature of the upper mantle-lower crust transition zone. Petrographic evidence shown in this study demonstrates, that in addition to porous flow, veins and dikes are crucial in extracting melt at mid-ocean ridges. This is supported by observations in mantle peridotite massifs (Nicolas, 1986, 1989) and by studies of physical properties of melts in veins and dikes under mantle conditions (Nicolas and Jackson, 1982; Sleep, 1988). In our samples from the South-East Pacific, we have documented a variety of textures and compositions of minerals crystallized in intrusive veins and under various mechanical and thermal conditions. These textures and compositions indicate, that magmatic infiltration and intrusion of basaltic melts originates in the upper mantle and continues up to the sub-sea floor, alimenting magma chambers at fast spreading centers. The occurrence of clinopyroxenite veins and plagioclase peridotites, and of numerous gabbroic veins in harzburgite provide direct evidence for the heterogeneous flow and channelization of melt through the upper mantle and lower crust. The overall effect is a local fertilization of the residual harzburgite.

With the ongoing study of new sites there is growing evidence that intrusion and percolation of basaltic melts in dikes and sills into host peridotites are much more common than previously thought, and that these processes occur not only at fast but also at slowly spreading ridges. Melt/rock interactions may lead to important mineralogical modifications, suggesting that this process has to be taken into account when modelling medium- to large-scale ridge segmentation based on sea-floor peridotite and basalt chemistry. Spinel and diopside compositions in harzburgite are strongly modified by magmatic intrusions resulting in large variations in their Fe, Ti, Cr, Mg, Al contents. It follows that caution is needed when using major element compositions to infer mantle conditions. In addition, trace element contents of both liquids and minerals could be modified as well. Dunites from the Terevaka transform fault are always associated with gabbroic intrusions and impregnations, suggesting that they are the product of melt/peridotite interactions and that they are not pure residues derived from large degrees of melting.

Acknowledgements

We are indebted to the Captain and crew of the R.V. Sonne (cruises 65 and 80). M. Bohn is kindly acknowledged for assistance during microprobe analyses and R. Apprioual for technical support. We are grateful to L. Reisberg and D. Bideau for comments, and to A. Nicolas and F. Boudier for reviews of the manuscript. We are also indebted to R.L.M. Vissers for his patience, and for many helpful comments during the final editing phase of the manuscript. M.C. is supported by a FCAR scholarship (Government of Quebec).

References

Bailey, J.C., Campsie, J., Hald, N., Dittmer, F. and Rasmussen, M., 1992. Petrology and geochemistry of a dredged clinopyroxenite-dolerite basal complex from the Jan Mayen volcanic province, Norwegian - Greenland sea. Mar. Geol., 105: 63-76.

Batiza, R. and Niu, Y., 1992. Petrology and magma chamber processes at the East Pacific Rise 9°30'N. J. Geophys. Res., 97: 6779-6797.

Bédard, J.H., 1992. Oceanic crust as a reactive filter: synkinematic intrusion, hybridization, and assimilation in an ophiolitic magma chamber, western Newfoundland. Geology, 21: 77-80.

Bédard, J.H. and Constantin, M., 1991. Syn- and post-kinematic intrusions of gabbros and peridotite into layered gabbroic cumulates in the Bay of Islands ophiolite Newfoundland: genesis of anorthosite by reaction and troctolite by hybridization. In: Current Research, Geol. Survey Canada, 91-1: 79-88.

Benn, K. and Allard, B., 1989. Preferred mineral orientations related to magmatic flow in ophiolite layered gabbros. J. Petrol., 30: 925-946.

Benn, K., Nicolas, A. and Reuber, I., 1988. Mantle-crust transition zone and origin of wehrlitic magmas: evidence from the Oman ophiolite. Tectonophysics, 151: 75-85.

Berger, E.T. and Vannier, M., 1984. Les dunites en enclaves dans les basaltes alcalins des iles oceaniques: approche pétrologique. Bull. Minéral., 107: 649-663.

Bideau, D., and Hékinian, R., 1994. A dynamic model for generating small-scale heterogeneities in ocean floor basalts. Submitted to J. Geophys. Res.

Bloomer, S.H., Natland, J.H. and Fisher, R.L. 1989. Mineral relationships in gabbroic rocks from fractures zones of Indian Ocean Ridges: evidence for extensive fractionation, parental diversity, and boundary-layer recrystallization. In: A.D. Saunders and M.J. Norry (Editors), Magmatism in the Ocean Basins, Geol. Soc. London Spec. Publ., 42: 107-124.

Bonatti, E., Peyve, A., Kepezhinskas, P., Kurentsova, N., Seyler, M., Skolotnev, S. and Udsintev, G., 1992. Upper mantle heterogeneity below the Mid-Atlantic ridge, 0°-15°N. J. Geophys. Res., 97: 4461-4476.

Boudier, F., 1991. Olivine xenocrysts in picritic magmas. Contrib. Mineral. Petrol., 109: 114-123.

Boudier, F. and Nicolas, A., 1972. Fusion partielle gabbroique dans la lherzolite de Lanzo. Bull. Suisse Minéral. Pétrol., 52: 39-56.

Boudier, F. and Nicolas, A., 1985. Harzburgite and lherzolite subtypes in ophiolitic and oceanic environments. Earth Planet. Sci. Lett., 76: 84-92.

Cannat, M., 1993. Emplacement of mantle rock in the seafloor at mid-ocean ridges. J. Geophys. Res., 98: 4163-4172.

Cannat, M., Bideau, D. and Hébert, R., 1990. Plastic deformation and magmatic impregnation in serpentinized ultramafic rocks from the Garrett transform fault (East Pacific Rise). Earth Planet. Sci. Lett., 101: 216-232.

Cannat, M., Bideau, D. and Bougault, H., 1992. Serpentinized peridotites and gabbros in the Mid-Atlantic Ridge axial valley at 15°37'N and 16°52'N. Earth Planet. Sci. Lett., 109: 87-106.

Ceuleneer, G. and Rabinowicz, M., 1992. Mantle flow and melt migration beneath oceanic ridges: models derived from observations in ophiolites. In: J. Phipps Morgan, D. Blackman and J.M. Sinton (Editors), Mantle flow and melt generation at Mid-Ocean ridges. Geophys. Monogr. 71: 123-154.
Constantin, M., 1992. Petrologie des roches gabbroiques du sondage 735B de la zone de fracture Atlantis II, dorsale sud-ouest Indienne, Océan Indien. M. Sc.Thesis, Univ. Laval, Québec: 200 pp.
Constantin, M. and Hébert, R., 1992. Petrology and geochemistry of gabbroic rocks from Atlantis II fracture zone, South-west Indian ridge: implications for multiple crustal oceanic processes., EOS Trans., Amer. Geophys. Union, 73: 359.
Constantin, M. and Hékinian, R., 1993. Les péridotites à plagioclase du Sud-Est Pacifique: témoins de l'intéraction des magmas basaltiques avec le manteau résiduel océanique. Abstract with program, Journées Spéc. de la Soc. Géol. de France (Géosc. Mar.), Paris: 56.
Constantin, M., Hékinian, R., Ackermand, D., Stoffers, P. and Francheteau, J., 1993. Upper mantle and lower crust exposed in the Easter microplate (South East Pacific). Terra Nova, V.5, Abstract suppl. no. 1, EUG VII: 184-185.
DeMets, C., Gordon, R.G., Argus, D.F. and Stein, S., 1990. Current plate motions. Geophys. J. Intern., 101: 425-478.
Dick, H.J.B., 1989. Abyssal peridotites, very slow spreading ridges and ocean ridge magmatism. In: A.D. Saunders and M.J. Norry (Editors), Magmatism in the Ocean Basins, Geol. Soc. London Spec. Publ., 42: 71-105.
Dick, H.J.B. and Bullen, T., 1984. Chromian spinel as a petrogenetic indicator in abyssal and alpine-type peridotites and spatially associated lavas. Contrib. Mineral. Petrol., 86: 54-76.
Dick, H.J.B., Meyer, P.S., Bloomer, S.H., Kirby, S., Stakes, D. and Mawer, C., 1991. Lithostratigraphic evolution of an in-situ section of oceanic layer 3. In: R.P. Von Herzen, P.T. Robinson et al. (Editors), Proc. of the ODP, Sci. Res., 118: 439-538.
Elthon, D., 1992. Chemical trends in abyssal peridotites: refertilization of depleted suboceanic mantle. J. Geophys. Res., 97: 9015-9025.
Elthon, D., Stewart, M. and Ross, D.K., 1992. Compositional trends of mineral in oceanic cumulates. J. Geophys. Res., 97: 15189-15199.
Engel, C.G. and Fisher, R.L., 1975. Granitic to ultramafic rock complexes of the Indian Ocean ridge system, western Indian Ocean. Geol. Soc. Am. Bull., 86: 1553-1578.
Evans, C., 1985. Magmatic metasomatism in peridotites from the Zambales ophiolite. Geology, 13: 166-169.
Fisher, R.L., Dick, H.J.B., Natland, J.H. and Meyer, P.S., 1986. Mafic/ultramafic suites of the slowly spreading Southwest Indian ridge: PROTEA exploration of the Antartica plate boundary, 54oE-47oE, 1984. Ofioliti, 11: 147-178.
Fontignie, D. and Schilling, J.-G., 1991. 87Sr/86Sr and REE variations along the Easter Microplate boundaries (south Pacific): application of multivariate statistical analyses to ridge segmentation. Chem. Geol., 89: 209-241.
Fox, P.J. and Gallo, D.G., 1984. A tectonic model for ridge-transform-ridge plate boundaries: implications for the structure of oceanic lithosphere. Tectonophysics, 104: 205-242.
Francheteau, J. and Ballard, R., 1983. The East Pacific Rise near 21°N, 13°N and 20°S: inferences for along-strike variability of axial processes of the Mid-Ocean Ridge. Earth Planet. Sci. Lett., 64: 93-116.
Francheteau, J. et al. (Rapanui Scientific Party), 1988. Pito and Orongo fracture zones: the northern and southern boundaries of the Easter microplate (southeast Pacific). Earth Planet. Sci. Lett., 89: 363-374.
Francheteau, J., Armijo, R., Cheminée, J.L., Hékinian, R., Lonsdale, P. and Blum, N., 1990. 1 Ma East Pacific Rise oceanic crust and uppermost mantle exposed by rifting in Hess Deep (equatorial Pacific Ocean). Earth Planet. Sci. Lett., 101: 281-295.
Girardeau, J. and Francheteau, J., 1993. Plagioclase-wehrlites and peridotites on the East Pacific Rise (Hess Deep) and the Mid-Atlantic Ridge (DSDP Site 334): evidence for magma percolation in the oceanic upper mantle. Earth Planet. Sci. Lett., 115: 137-149.
Hamlyn, P.R. and Bonatti, E., 1980. Petrology of mantle-derived ultramafics from the Owen fracture zone, northwest Indian ocean: implications for the nature of the oceanic upper mantle. Earth Planet. Sci. Lett., 48: 65-79.
Hanan, B.B. and Schilling, J.-G., 1989. Easter microplate evolution: Pb isotope evidence., J. Geophys. Res., 94: 7432-48.
Harte, B., Hunter, R.H. and Kinny, P.D., 1993. Melt geometry, movement and crystallization, in relation to mantle dykes, veins and metasomatism. Phil. Trans. R. Soc. Lond., A342: 1-21.
Hébert, R. and Constantin, M., 1991. Petrology of hydrothermal metamorphism of oceanic layer 3: implications for sulfide paragenesis and redistribution., Econ. Geol., 86: 472-485.
Hébert, R., Bideau, D. and Hékinian, R., 1983. Ultramafic and mafic rocks from the Garrett transform fault near 13°30'S on the East Pacific Rise: igneous petrology. Earth Planet. Sci. Lett., 65: 107-125.

Hébert, R., Serri, G. and Hékinian, R., 1989. Mineral chemistry of ultramafic tectonites and ultramafic to gabbroic cumulates from the major oceanic basins and Northern Apennine ophiolites (Italy) - A comparison. Chem. Geol., 77: 183-207.
Hébert, R., Constantin, M. and Robinson, P.T., 1991. Primary mineralogy of Leg 118 gabbroic rocks and their place in the oceanic spectrum of oceanic mafic igneous rocks. In: R.P. Von Herzen, P.T. Robinson et al. (Editors), Proc. of the ODP, Sci. Res., 118: 3-20.
Hékinian, R., 1970. Gabbro and pyroxenite from a deep-sea core in the Indian Ocean. Mar. Geol., 9: 287-294.
Hékinian, R., Bideau, D., Cannat, M., Francheteau, J. and Hébert, R., 1992. Volcanic activity and crust-mantle exposure in the ultrafast Garrett transform near 13°28'S in the Pacific. Earth Planet. Sci. Lett., 108: 259-275.
Hékinian, R., Bideau, D., Francheteau, J., Cheminée, J.-L., Armijo, R., Lonsdale, P. and Blum, N., 1993. Petrology of the East Pacific Rise crust and upper mantle exposed in Hess Deep (Eastern equatorial Pacific). J. Geophys. Res., 98: 8069-8094.
Henderson, P., 1975. Reaction trends shown by chrome-spinels of the Rhum layered intrusion. Geochim. Cosmochim. Acta, 39: 1035-1044.
Henstock, T.J., Woods, A.W. and White R.S., 1993. The accretion of oceanic crust by episodic sill intrusion. J. Geophys. Res., 98: 4143-4161.
Hey, R.N., Naar, D.F., Kleinrock, M.C., Phipps Morgan, W.J., Morales, E. and Schilling, J.-G., 1985. Microplate tectonics along a superfast seafloor spreading system near Easter Island. Nature, 317: 320-325.
Hirose, K. and Kushiro, I., 1993. Partial melting of dry peridotites at high pressures: determination of compositions of melts segregated from peridotite using aggregates of diamond. Earth Planet. Sci. Lett., 114: 477-489.
Ildefonse, B., Nicolas, A. and Boudier, F., 1993. Evidence from the Oman ophiolite for sudden stress changes during melt injection at oceanic spreading centres. Nature, 366: 673-675.
Johnson, K.T.M., Dick, H.J.B. and Shimizu, N., 1990. Melting in the oceanic upper mantle: an ion microprobe study of diopsides in abyssal peridotites., J. Geophys. Res., 95: 2661-2678.
Karson, J.A. and Dick, H.J.B., 1984. Deformed and metamorphosed oceanic crust on the Mid-Atlantic ridge. Ofioliti, 9: 279-302.
Kelemen, P.B., 1990. Reaction between ultramafic rock and fractionating basaltic magma - I. Phase relations, the origin of calc-alkaline magma series, and the formation of discordant dunite. J. Petrol., 31: 51-98.
Kelemen, P.B., Dick, H.J.B. and Quick, J.E., 1992. Formation of harzburgite by pervasive melt/rock reaction in the upper mantle. Nature, 358: 635-641.
Kelemen, P.B., Joyce, D.B., Webster, J.D. and Holloway, J.R., 1990. Reaction between ultramafic rock and fractionating basaltic magma - II. Experimental investigation of reaction between olivine tholeiite and harzburgite at 1150-1050°C and 5kb. J. Petrol., 31: 99-134.
Klein, E.M. and Langmuir, C.H., 1987. Global correlations of ocean ridge basalt chemistry with axial depth and crustal thickness. J. Geophys. Res., 92: 8089-8115.
Kornprobst, J., Ohnenstetter, D. and Ohnenstetter, M., 1981. Na and Cr contents in clinopyroxenes from peridotites: a possible discriminant between sub-continental and sub-oceanic mantle. Earth Planet. Sci. Lett., 53: 241-254.
Langmuir, C.H. and Bender, J.F., 1984. The geochemistry of oceanic basalts in the vicinity of transform faults: observations and implications. Earth Planet. Sci. Lett., 69: 107-127.
Lykins, R.W. and Jenkins, D.M., 1992. Experimental determination of pargasite stability relations in the presence of orthopyroxene. Contrib. Mineral. Petrol., 112: 405-413.
Martinez, F., Naar, D.F., Reed, T.B. and Hey, R.N., 1991. Three-dimensional SeaMARC II, gravity, and magnetics study of large-offset rift propagation at the Pito rift, Easter microplate. Mar. Geophys. Res., 13: 255-285.
Mével, C., 1987. Evolution of oceanic gabbros from DSDP Leg 82: influence of the fluid phase on metamorphic crystallizations. Earth Planet. Sci. Lett., 83: 67-79.
Mével, C., Cannat, M., Gente, P., Marion, E., Auzende, J.M. and Karson, J.A., 1991. Emplacement of deep crustal and mantle rocks on the west median valley wall of the MARK area (MAR, 23°N). Tectonophysics, 190: 31-53.
Michael, P.J. and Bonatti, E., 1985. Peridotite composition from the North Atlantic: regional and tectonic variations and implications for partial melting. Earth Planet. Sci. Lett., 73: 91-104.
Naar, D.F. and Hey, R.N., 1989. Recent Pacific-Easter-Nazca plate motions. In: Sinton, J.M. (Ed.), Evolution of Mid Ocean Ridges. Geophys. Monogr. 57, IUGG-AGU, 8: 9-30.
Naar, D.F. and Hey, R.N., 1991. Tectonic evolution of the Easter microplate. J. Geophys. Res., 96: 7961-

7993.
Naar, D.F., Martinez, F., Hey, R.N., Reed, T.B. and Stein, S., 1991. Pito rift: how a large-offset rift propagates. Mar. Geophys. Res., 13: 287-309.
Natland, J.H., 1980. Effect of axial magma chambers beneath spreading centers on the compositions of basaltic rocks. In: B.R. Rosendahl, R. Hékinian et al. (Editors), Init. Rep. of DSDP, 54, Washington: 833-850.
Nicholls, I.A., Ferguson, J., Jones, H., Marks, G.P. and Mutter, J.C., 1981. Ultramafic blocks from the ocean floor southwest of Australia. Earth Planet. Sci. Lett., 56: 362-374.
Nicolas, A., 1986. A melt extraction model based on structural studies in mantle peridotites. J. Petrol., 27: 999-1022.
Nicolas, A., 1989. Structures of ophiolites and dynamics of oceanic lithosphere. Dordrecht, Kluwer: 367 pp.
Nicolas, A., 1992. Kinematics in magmatics rocks with special reference to gabbros. J. Petrol., 33: 891-915.
Nicolas, A. and Jackson, M., 1982. High temperature dikes in peridotites: origin by hydraulic fracturing. J. Petrol., 23: 568-582.
Nicolas, A. and Prinzhofer, A., 1983. Cumulative or residual origin for the transition zone in ophiolites: structural evidence. J. Petrol., 24: 188-206.
Phipps Morgan, J. and Chen, Y.J., 1993. The genesis of oceanic crust: magma injection, hydrothermal circulation, and crustal flow. J. Geophys. Res., 98: 6283-6297.
Poreda, R.J., Schilling, J.G., and Craig, H., 1993. Helium isotope ratios in Easter microplate basalts. Earth Planet. Sci. Lett., 119: 319-329.
Quick, J.E., 1981. The origin and significance of large, tabular dunite bodies in the Trinity peridotite, Northern California. Contrib. Mineral. Petrol., 78: 413-422.
Quick, J.E. and Delinger, R.P., 1993. Ductile deformation and the origin of layered gabbro in ophiolites. J. Geophys. Res., 98: 14015-14027.
Rampone, E., Piccardo, G.B., Vannucci, R., Bottazzi, P. and Ottolini, L., 1993. Subsolidus reactions monitered by trace element partitioning: the spinel- to plagioclase-facies transition in mantle peridotites. Contrib. Mineral. Petrol., 115: 1-17.
Reuber, I., Whitechurch, H. and Juteau, T., 1985. Successive generations of coarse grained dikelets in the ophiolite complex of Antalya, Turkey: products of partial fusion and residual liquids. Ofioliti, 10: 35-62.
Ridley, W.I., 1977. The crystallization trends of spinels in tertiary basalts from Rhum and Muck and their petrogenetic significance. Contrib. Mineral. Petrol., 64: 243-255.
Roeder, P.L. and Reynolds, I., 1991. Crystallization of chromite and chromium solubility in basaltic melts. J. Petrol., 32: 909-934.
Rusby, R.I., 1992. GLORIA and other geophysical studies of the tectonic pattern and history of the Easter Microplate, southeast Pacific. In: L.M. Parson, B.J. Murton and P. Browning (Editors), Ophiolites and their Modern Oceanic Analogus, Geol. Soc. London Spec. Publ., 60: 81-106.
Schilling, J.-G., Sigurdsson, H., Davis, A.N. and Hey, R.N., 1985. Easter microplate evolution. Nature, 317: 325-331.
Searle, R.C., Rusby, R.I., Engeln, J., Hey, R.N., Zukin, J., Hunter, P.M., LeBas, T.P., Hoffman, H.-J. and Livermore, R., 1989. Comprehensive sonar imaging of the Easter microplate. Nature, 341: 701-705.
Searle, R.C., Bird, R.C., Rusby, R.I. and Naar, D.F., 1993. The development of two oceanic microplates: Easter and Juan Fernandez microplates, East Pacific Rise. J. Geol. Soc. London, 150: 965-976.
Shibata, T. and Thompson, G., 1986. Peridotites from the Mid-Atlantic ridge at 43°N and their petrogenetic relation to abyssal tholeiites. Contrib. Mineral. Petrol., 93: 144-159.
Sinton, J.M. and Detrick, R.S., 1992. Mid-ocean ridge magma chambers. J. Geophys. Res., 97: 197-216.
Sleep, N.H., 1988. Tapping of melt by veins and dikes. J. Geophys. Res., 93: 10255-10272.
Solomon, S.C. and Toomey, D.R., 1992. The structure of mid-ocean ridges. Ann. Rev. Earth Planet. Sci., 20: 329-364.
Stoffers, P., Hékinian, R. et al., 1989. Cruise Report SONNE 65 - Midplate II, Hotspot volcanism in the central Southpacific. Berichte - Reports, no.40, Univ. Kiel: 126 pp.
Stoffers, P., Hékinian, R. et al., 1992. Cruise Report SONNE 80a - Midplate III, Oceanic volcanism in the Southpacific. Berichte - Reports, no.58, Univ. Kiel: 128 pp.
Takahashi, E., Shimazaki, T., Tsuzaki, Y. and Yoshida, H., 1993. Melting study of a peridotite KLB-1 to 6.5GPa, and the origin of basaltic magmas. Phil. Trans. R. Soc. Lond., A342: 105-120.
White, R.S., McKenzie, D. and O'Nions, R.K., 1992. Oceanic crustal thickness from seismic measurements and rare earth element inversions. J. Geophys. Res., 97: 19683-19715.
Wilshire, H.G. and Kirby, S.H. 1989. Dikes, joints, and faults in the upper mantle. Tectonophysics, 161: 23-31.

Part II

Ophiolite Studies

Plastic Deformation of Gabbros in a Slow-spreading Mesozoic Ridge: Example of the Montgenèvre Ophiolite, Western Alps

R. CABY
Laboratoire de Tectonophysique, Université Montpellier II - CNRS, Place Bataillon, 34095 Montpellier Cedex 05, France.

Abstract

Gabbros of the Montgenèvre ophiolite (external Piemont zone, western Alps), with the mineralogical, geochemical and isotopic characteristics of an oceanic ridge, show evidence of pervasive high-temperature deformation before the intrusion of basalt dikes. This deformation is incipient in the layered troctolites immediately above a locally preserved paleo-Moho, and increases in intensity upward in the overlying olivine-poor gabbros producing an anastomosing system of mylonitic shear zones and several steep ultramylonite belts. The mylonites and ultramylonites developed by solid-state shearing of the gabbros at temperatures in excess of 800-850 °C, allowing recrystallization of augite subgrains in the wings of ductilely deformed magmatic diopside-salite clasts followed by crystallization of brown amphiboles. On the basis of textural relationships it is proposed that syn-kinematic partial melting within the sheared gabbros, at temperatures of 850° C or higher, generated augite- and pargasite-bearing leucodioritic magmas which evolved in narrow magmatic conduits and percolated through tectonites and actively deforming mylonite zones, to form cross-cutting veins in adjacent less deformed gabbros. The complex geometry of the structures and their kinematics, when restored to a stage before the onset of pre-Alpine and Alpine brittle tectonics, suggest that this tectono-metamorphic evolution started after rapid solidification of the gabbros. It is proposed that the gabbroic mass was pushed away from a spreading center due to the activity of extensional and transform shear zones, in response to spreading through continuous domal uplift of the underlying mantle peridotites from which, at deeper levels, new basaltic melt was extracted to produce off-axis basaltic volcanism. The "magmatic" Moho became thereby transformed into a "tectonic" moho, followed by uplift of peridotites to the ocean floor where they became overlain by ultramafic-derived sediments such as to form a "sedimentary" Moho. The scarce geochronological data suggest a long time span, of about 50-60 Ma, between magmatic crystallization of gabbros and diorites (212± 8 Ma) and the deposition of the oceanic cover (165-

R.L.M. Vissers and A. Nicolas (Eds.),
Mantle and Lower Crust Exposed in Oceanic Ridges and in Ophiolites, 123-145.

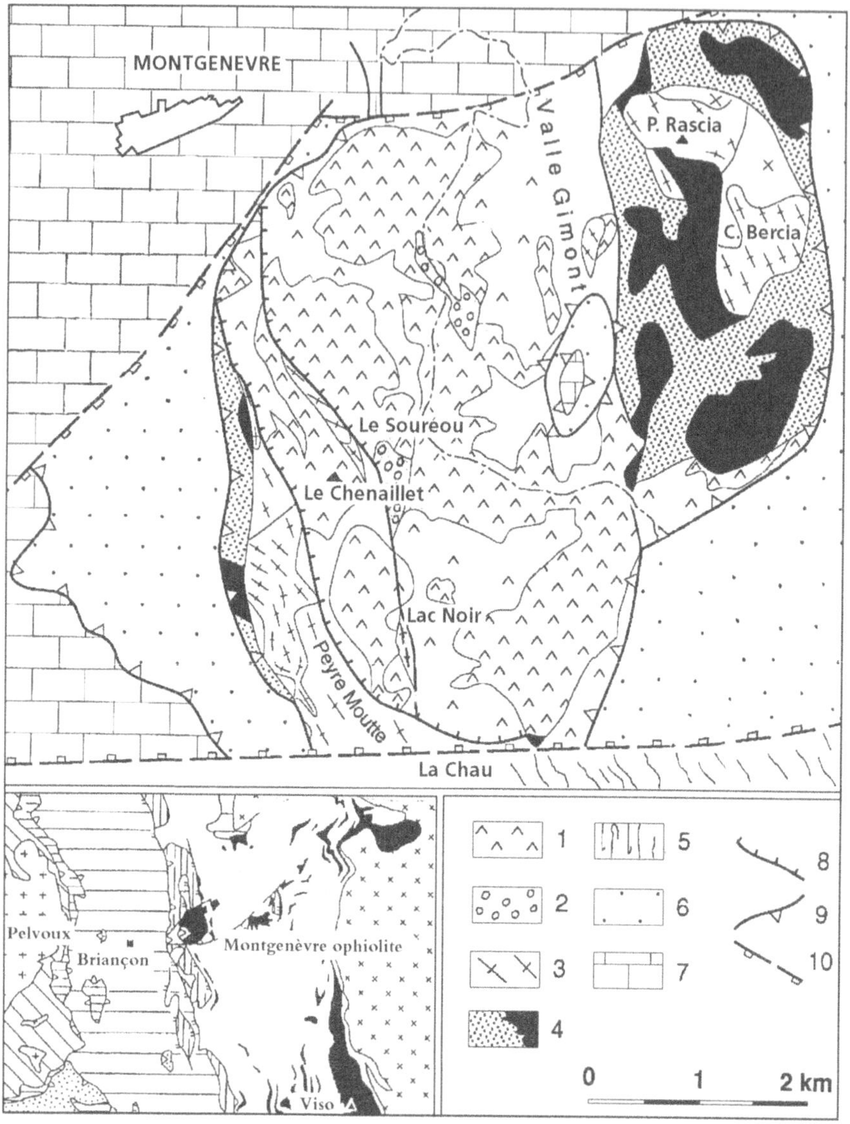

Figure 1. Simplified geological map of the Montgenèvre ophiolite (modified after Bertrand et al., 1987; minor faults not represented and Recent deposits removed). Ophiolite unit: (1) undifferentiated pillowed basalts, dolerite dikes not shown; (2) major outcrops of ultramafic arenites and conglomerates; (3) gabbros and dioritic rocks; (4) serpentinized peridotites, dotted where hidden below Recent terrains. Spaced decoration in 1 to 3 delineates assumed extension of units below Recent deposits. Oceanic cover sequences: (5) Upper Jurassic - Cretaceous Schistes lustrés with blueschist facies, undifferentiated; (6) Replatte-Lago Nero unit (idem, with lower grade metamorphism). Passive Tethyan paleo-margin: (7) Upper Triassic dolomites and lower Jurassic sediments; (8) Main Jurassic fault; (9) Alpine thrust; (10) late Alpine fault.

160 Ma), suggesting a slow-spreading paleoridge in the relatively narrow (<1000 km) Piemonte-Ligurian branch of the Tethys ocean.

Introduction

The ophiolitic massifs of the Piemonte-Ligurian zone of the western Alps show distinct characteristics suggesting that they are remnants of a narrow oceanic basin that may have formed by limited, oblique stretching and breakup of the continental lithosphere (Weissert and Bernoulli, 1985; Bernoulli and Lemoine, 1987; Lemoine et al., 1987). In this context it is important to note that the plagioclase lherzolites which are predominant in the Piemont zone display mineralogical and geochemical features indicative of a rather low degree of partial melting (Ishiwatari, 1985; Boudier and Nicolas, 1977; Bodinier, 1988). Gabbros constitute small massifs, mostly intruding lherzolites (Lombardo and Pognante, 1982) and no remnants of significant dike complexes have been reported. Moreover, it has in many localities been shown that exposure of the serpentinized lherzolites occurred prior to the pouring of submarine basalts and associated dolerite sills (Lemoine, 1980; Lemoine et al., 1987; Caby et al., 1987; Lagabrielle et al., 1984). As a consequence, structural and petrological studies of the western Alps ophiolites till date have failed to document a clearly defined paleo-Moho. Radiolarites were deposited onto basalts, gabbros and mantle peridotites between the Callovian and the middle Kimmeridgian (de Wever and Caby, 1981; de Wever et al., 1987). Taking into account the 212 ± 8 Ma fission-track age of magmatic zircons from the Montgenèvre gabbros (Carpena and Caby, 1984) considered as the minimum crystallization age of the gabbros, it follows that a long-lived, delayed oceanization > 60 Ma may have taken place in this part of the Tethys.

The Montgenèvre ophiolite (Fig. 1) is one of the best preserved massifs from the external Piemont Zone. Below I report new petrological and kinematic data from this massif including uncommon kinematic and petrological features associated with extensive high-temperature plastic deformation in the gabbros. Then I discuss their significance, to suggest low spreading rates in this part of the early Mesozoic, Piemonte-Ligurian Tethys.

Regional geology

The Montgenèvre ophiolite massif forms a thin klippe, which must be less than 1000 m thick as shown by the Valle Gimont tectonic window. This klippe of the frontal part of the Piemont zone rests onto the Lago Nero unit with blueschist facies mineral assemblages, itself overlying Triassic fig-dolomites (Figs. 1 and 2). The ophiolite is a remnant of a higher nappe that suffered minor Alpine ductile deformation and metamorphism and is now preserved in a down-faulted block. As pointed out by Blake and Jayko (1990), late-Alpine extension may have considerably thinned the former tectonic pile, as suggested by the juxtaposition of jadeite-glaucophane-lawsonite-bearing Schistes lustrés (with pure jadeite + quartz in the Cervières window) and the much lower grade oceanic assemblages of the klippe. Unfortunately, outcrops are often tilted by land slides, and many significant observations had to be made in large blocks from land slides, screes and moraines, especially those around Peyre Moutte.

The geometry of the western part of the klippe suggests a very reduced, right way-up section of mantle and oceanic crust (Figs. 1 and 2). Gabbros near Le Soureou and Le Lac Noir may represent a narrow linear extrusion that has pearced overlying basalts exposed in the central part of the klippe. The eastern part of the massif is bounded to the west by a complex fault zone adjacent to the Valle Gimont window. The Punta Rascia and Colle Bercia gabbros are interpreted as intrusive bodies emplaced in the mantle peridotites.

Petrological features

Peridotites

Unlike most other ophiolitic massifs of the Piemont zone essentially dominated by plagioclase lherzolites (Nicolas et al., 1971; Boudier, 1972; Lemoine et al., 1987), the Montgenèvre ophiolite mainly comprises spinel harzburgites and lherzolites, typically with 1-2% CaO, 35-37% MgO and 0.2-03% Cr_2O_3, less plagioclase lherzolites and minor wehrlites and dunites with cumulate characteristics (Bertrand et al., 1981, 1982; Bertrand et al., 1987). The amount of serpentinization generally exceeds 80%. Less serpentinized samples display coarse-grained textures, with orthopyroxene clasts of commonly 1 to 2 cm length. These pyroxenes have a shape fabric defining a foliation which, in many localities, cuts at moderate angles

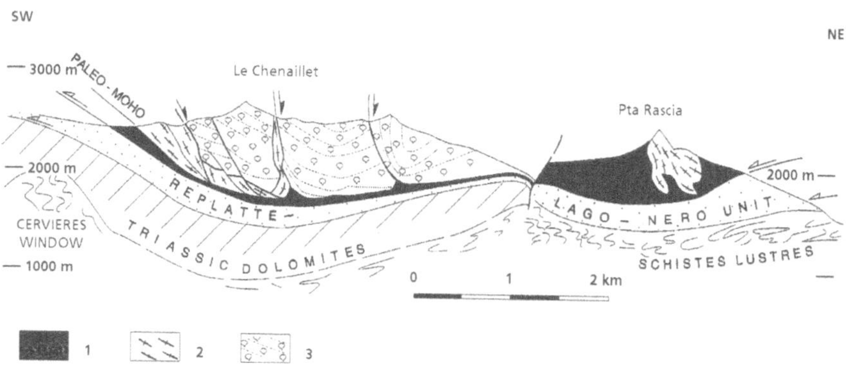

Figure 2. Geological cross-section across the Montgenèvre ophiolite klippe: (1) mantle peridotites; (2) gabbros; (3) basalts. Note that section differs from that of Blake and Jayko (1990) in which oceanic tectonics is not taken into acount.

across the compositional layering. In addition, there are few coarse grained (> 3 cm) pyroxenite veins, 5 to 10 centimeters thick, with an incipient foliation. In more deformed rock types, the orthopyroxene clasts tend to display sub-spherical shapes wrapped by a strongly curved foliation, whereas other clasts form thin ribbons. Clinopyroxene clasts are recrystallized along their rims and in adjacent pressure shadows into a mosaic of secondary clinopyroxene. In mylonitic samples, serpentinized clasts form thin (< 1 mm) ribbons, up to several centimeters long, whilst fragmented and needle-like Cr-spinels show incipient replacement by plagioclase invariably altered into hydrogarnet and chlorite. One rather exceptional sample (1464, collected from the Peyre Moutte moraine) contains extremely stretched orthopyroxene clasts, about one millimeter thick and several centimeters long, associated with ribbons of finely crushed and unrecrystallized sub-grains of fresh olivine, alternating with polycrystalline ribbons of brown-red pargasitic amphibole partly overgrown by tremolite, serpentine and talc (see Fig. 5a below). Such a mylonitic microstructure suggests that this rock originates from a crustal shear zone, in which the primary mantle fabric and assemblages were reworked and strongly altered by fluids under amphibolite to granulite facies conditions. The inferred shear zone may have been active coeval with the high-temperature deformation in the gabbros documented below.

Two localities (4 km NW of Le Chenaillet and south of Punta Rascia) allow to measure the orientation of the mantle foliation which dips 45° to the SE, i.e., parallel to the inferred paleo-Moho, both with a moderately dipping stretching lineation trending ENE. The cumulate rocks (pyroxenite, dunite and wehrlite) are free of post-magmatic deformation. They represent lenses and pockets intruded in mantle peridotites (Bertrand et al., 1981; 1982).

Magmatic features of gabbroic rocks

Layered troctolites and olivine gabbros (less than 100 m) are coarse-grained rocks with a locally well-preserved, centimeter- to decimeter-scale layering defined by a variable abundance of plagioclase. These rocks are exposed at the base of the southwestern ridge of Le Chenaillet and occur as blocks in related land slides to the south. Totally altered olivine is occasionally rimmed by or enclosed in clinopyroxene, both minerals being enclosed in plagioclase which is mostly altered. Late magmatic rims of pargasitic amphibole are also present around ilmenite. Anorthosites make up both concordant layers and cross-cutting veins and some of these rocks are enriched in ilmenite.

Near the base (< 50 m) of the massif, all of these layered rocks are nearly free of post-magmatic, high-temperature deformation. They are cut by mafic ferrogabbro to pyroxenite veins, one to some tens of centimeters

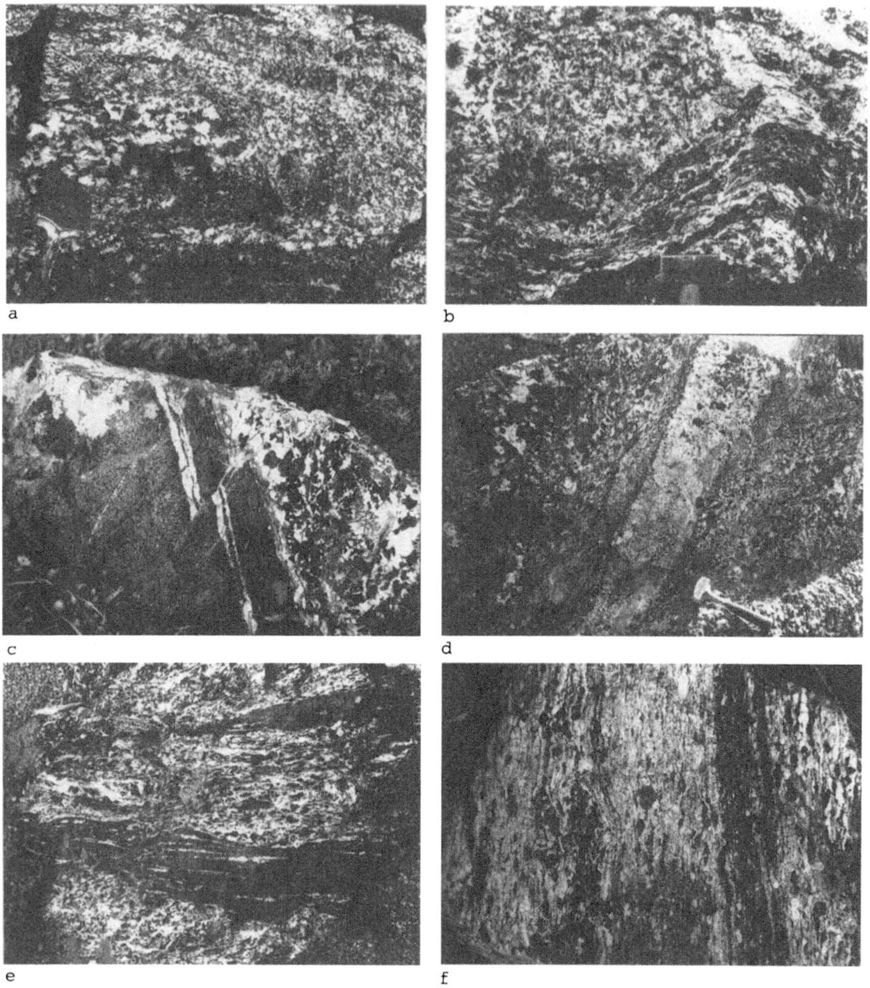

Figure 3. Progressive ductile deformation of gabbros and diorites: (a) poorly layered clinopyroxene gabbros nearly free of solid-state deformation; (b) blocks of undeformed gabbro cut by anastomosing shear zones; (c) sharp gradient of deformation between undeformed, pegmatitic gabbro and amphibolite with plagioclase ribbbons; latter rock represents margin of diorite dike with magmatic microstructure in central part only, area of view about 1 x 0.50 m^2; (d) undeformed pegmatitic leucodiorite vein, cutting coarse-grained clinopyroxene gabbro affected by solid-state deformaton which increases towards dike walls; (e) amphibolite lenses derived from sheared mafic diorite dikes, showing sharp contacts with adjacent protomylonitic gabbro; (f) banded leucodiorite and mafic diorite alternating with protomylonitic gabbro (area of view appr. 15 x 25 cm^2).

thick, deeply altered by low-temperature ocean-floor metamorphism producing chlorite, serpentine, deep green Cl-bearing amphibole, sphene and epidote, and enriched in Fe-sulphides. Relics of large clinopyroxene, deep-brown amphibole, magnetite, and up to one centimeter long acicular apatite (samples with P_2O_5 up to 2.5%) are occasionally preserved, and allow to recognize a derivation from coarse-grained rock types.

The inferred gently east-dipping contact, covered by scree and moraines, between this lower sequence and the underlying coarse-grained spinel lherzolites exposed southwest of this crest is interpreted as a fragment of a paleo-Moho.

Clinopyroxene, olivine-poor gabbros and associated rocks (<300 m) overly olivine gabbros with some recurrences of Mg types. They form a poorly layered sequence made up of a patchwork of undeformed, coarse-grained and fine-grained rocks with subordinate pegmatite patches and pockets of microgabbro (Fig. 3a). The rocks are affected by a system of anastomosing shear zones marked by the development of a pronounced solid-state foliation (Fig. 3b) discussed in detail below. Clinopyroxene and plagioclase are the major constituents of these rocks, with minor olivine and opaques. Clinopyroxenes of the diopside-salite type show variable stages of replacement and are overgrown by late-magmatic, brown amphiboles with magnesio-hornblende compositions (Mével et al., 1978), which in turn are replaced and/or rimmed by green hornblende. Calcic plagioclase is mostly replaced by a cryptocrystalline assemblage of zoisite and clinozoisite + albite (Bertrand et al., 1981, 1982).

Fine-grained gabbros, mostly free of high-temperature deformation, form diffuse patches made up of sub- to euhedral grains of augite, plagioclase, altered olivine, and titano-magnetite. Pods of pegmatitic gabbro with clinopyroxenes of up to 20 centimeters are also frequent. Fe-Ti-rich, coarse-grained gabbros (with Fe_2O_3 up to 20%, TiO_2 up to 6.5%, and P_2O_5 up to 2.5%) form lenses always affected by a pervasive, late- to post-magmatic, high-temperature deformation and grade into amphibolites.

According to Bertrand et al. (1987), these gabbros show a wide range in element concentrations and result from differentiation processes different from basalts. Their low K_2O values (0.01<0.1) and high ε Nd in the range of +8 to 10 (S. Costa and C. Pin, written communication) confirms their oceanic character.

Dioritic rocks

Dioritic rocks occur both as veinlets, veins and dikes with either diffuse or sharp boundaries (e.g., Fig. 3c and d), and as diffuse impregnations in sheared gabbros (Fig. 4a, b and c). They are more abundant in the upper part of the gabbro sequence. In places, dioritic rocks make up about 20% of the rock volume.

With progressive deformation, these differentiated rocks grade into high-temperature protomylonites and mylonites (Fig. 3e and f) but never into ultramylonites.

The primary igneous mineralogy of the dioritic rocks includes oligoclase, colorless augite, brown amphibole of pargasitic composition, and abundant accessory minerals including ilmenite, sphene, apatite and U-

Figure 4. Mylonites and dioritic rocks (sections parallel to XZ, scale bars 3 cm): (a) fine-grained diorite in extensional shear bands and infiltrating along foliation planes of sheared gabbro; (b) veinlet of fine-grained augite diorite possibly generated by partial melting of mylonitic gabbro; dark ribbons consist of secondary augite, dark clasts are diopside-salite; note mineral zoning at base of vein; (c) patchy late-kinematic replacement of gabbro mylonitic fabric by fine-grained dioritic rock; (d) composite veinlet of fine-grained diorite and porphyritic leucodiorite cutting foliated gabbro; (e) folded mylonitic foliation in mylonitic gabbro, and adjacent ultramylonite, in core of major shear zone from Le Soureou; eye-shaped clinopyroxene clasts and ribbons of former plagioclase are surrounded by fine-grained, pargasite-rich dioritic matrix; (f) mylonite band cutting heterogeneous gabbro; dark rock within core of shear zone is a fine-grained diorite affected by moderate late-magmatic deformation;

poor zircon (Fig. 5e and f). Disseminated clasts of a more calcic plagioclase and of diopside-salite, both incorporated from the adjacent gabbros, are also present. Both mafic diorites and leucodiorites are quartz-free, K_2O-depleted (< 0.01).

Augite- and amphibole-bearing diorite veinlets decorate extensional

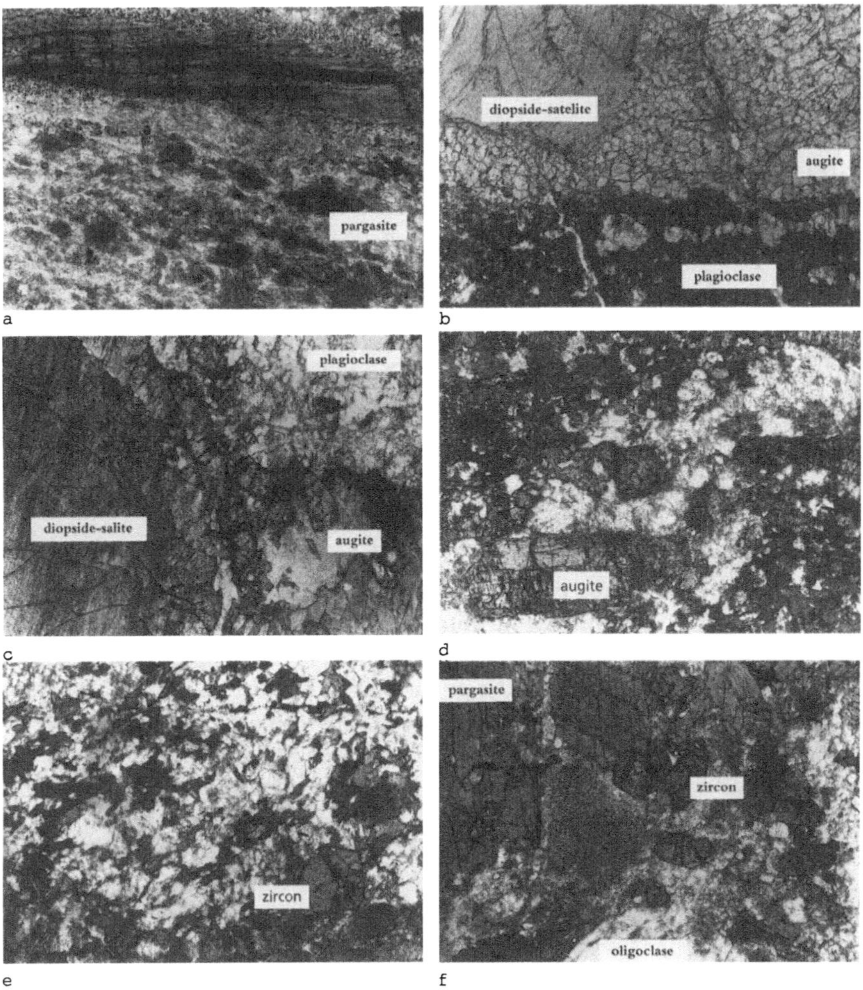

Figure 5. Microstuctures of mylonites and dioritic rocks in plane polarized light, area of view 2 x 3 cm^2: (a) part of orthopyroxene clast (upper third), more than 5 cm long, in fine-grained matrix of fresh pargasite; (b) strained diopside-salite clast in anhydrous mylonitic gabbro, recrystallized into fine-grained secondary augite; note recrystallized cloudy plagioclase in lower half. (c) clinopyroxene clast rimmed by secondary augite progressively incorporated into surrounding diorite; (d) anhydrous dioritic veinlet with euhedral augite cutting mafic amphibolite; (e) fine-grained diorite with recrystallized oligoclase and poikilitic pargasite; note euhedral zircon with plagioclase and amphibole inclusions; (f) coarse-grained diorite with idiomorphic pargasite and zircon phenocrysts; note recrystallized oligoclase.

shears in deformed gabbros (Fig. 4a). They impregnate and infiltrate along the foliation plane and progressively replace most of the sheared to mylonitic microstructure in the gabbro leading to a mylonitic ghost structure defined by the disposition and shape fabric of disseminated pyroxene and olivine porphyroclasts (Fig. 4c). Close to pyroxene clasts, leucodiorite veinlets are richer in magmatic augite that also mantles the clasts (Fig. 5c).

Other dioritic veinlets with a zoned texture clearly root in highly sheared gabbro and expand upwards to feed diffuse lenses (Fig. 4b). Such textural relationships strongly suggest that the leucodioritic magma formed by partial melting of the gabbro.

The frequently poikilitic habit of igneous minerals with a clear zoning, such as brown amphibole and oligoclase with a more calcic xenocryst core, and the amount of partly dissolved clasts also point to the in situcrystallization of these diffuse veinlets (Fig 5e, f). Accumulation of brown amphibole as patches and veins replacing or infiltrating the deformed texture of pyroxene-rich layers in ferrogabbro also suggests fluid transfer associated with the crystallizing dioritic melt.

Diorite dikes, when free of syn-kinematic recrystallization, are frequently composite, with alternating screens of mesocratic and leucocratic rocks (Fig. 3c). Others show a breccia texture. Some of the dikes were clearly emplaced within already formed shear zones, whilst veinlets of fine-grained isotropic diorite cut, with a diffuse margin, across deformed gabbro (Fig. 4d) or even diorite. Mafic diorite veins have also been seen cutting leucodiorite dikes.

Several diorite dikes contain trails of megacrysts of oligoclase locally packed together and, in smaller quantities, also of amphibole. Both types of megacrysts are up to several centimeters in size and embedded in fine-grained mesocratic diorite. Such patches are enriched in euhedral apatite and zircon of up to 3 millimeters, the latter mineral being frequently poikilitic with inclusions of euhedral oligoclase and brown amphibole (Fig. 5f). Fine-grained diorites occasionally preserve a magmatic structure defined by the alignment of euhedral plagioclase laths with inclusions of acicular brown amphibole. These magmatic rocks pass, over a distance of a few centimeters, into deformed equivalents with a metamorphic texture and a foliation defined by a shape preferred orientation of amphiboles and ilmenite trails. In thin section, eye-shaped former oligoclase phenocrysts contain strongly bent twinning planes, their margins are recrystallized into a mosaic microstructure, and adjacent pressure shadows are filled with a fine-grained assemblage of brown amphibole, ilmenite and plagioclase. In fact, at the scale of a few centimeters in a single dike, all gradations exist between igneous and metamorphic textures in an essentially similar mineralogy. Some sheared dikes contain younger pods of undeformed leucocratic diorite and albitite veinlets with an agmatitic structure. These pods and veinlets contain crystals of zoned euhedral plagioclase with oligoclase cores mantled by voluminous albite, megacrysts of poikilitic pale-brown to green amphibole, and late sphene.

Pegmatitic diorite dikes were mostly emplaced in the less deformed gabbros (Fig. 3d). They have amphibole-enriched margins and may contain angular blocks with corroded and amphibolitized rims. Patches of pegmatite in the basal troctolites contain amphibole and oligoclase crystals, up to 10 centimeters long, grown perpendicular to the vein walls.

On the basis of the above observations I conclude that the dioritic magmas were affected by complex phenomena of crystal accumulation in small chambers. They were emplaced repeatedly during high-temperature shearing of consolidated gabbro and possibly originate from partial hydrous melting and/or active leaching of the gabbros during shearing under hydrous conditions, possibly because sea water accessed shear zones and emergent faults. The lack of evidence for the usual crystallization sequence in leucodioritic melts (i.e., late poikilitic zircon) suggests a long-lived interaction between crystals and liquid, possibly related to continuous partial melting and/or leaching of gabbros during shearing. Alternatively, the dioritic melts may represent a residual liquid trapped in shear zones during high-temperature deformation, but its composition is different from the usual quartz-saturated, Na-rich residual liquids of oceanic tholeiitic magmas.

Albitites

Dikes and sills of albitite were emplaced in peridotites and gabbros and, according to Chapelle (1990), even in basalts. They are quartz-free, hololeucocratic rocks which in places enclose angular blocks of more mafic rock types, occasionally with cuspate contacts with the surrounding albitite. They typically contain more than 90% of albite with a fine-grained igneous microstructure of randomly oriented grains. Accessory minerals include tremolite-actinolite pseudomorphs of a magmatic amphibole, opaques, zircon, apatite and allanite. Associated brecciated mafic differentiates contain primary brown amphibole. The close association in some dikes of albitite with leucodiorite may indicate that the albitite represents the more evolved liquids differentiated from a dioritic magma. From geochemical considerations and using the typological classification of zircons of Pupin (1980), Chapelle (1990) has concluded that they may represent immiscible, mantle-derived magmas of tholeitic character that suffered fractional crystallization.

Volcanic rocks

Most basalts display spectacular pillow structures in sequences several tens of meters thick, as well as gradual transitions from massive flows to brecciated facies. Intercalated in the basalts occur layers of pillow breccia, hyaloclastite and minor volcanoclastic sediment. From a geochemical point of view, the basalts are typical MORBs with a narrow range in composition, suggesting a simple differentiation trend controlled by olivine and plagioclase fractionation (Bertrand et al., 1987). Basalt dikes without chilled margins cut deformed gabbros (Fig. 6) and mantle peridotites. Most of these dikes are several tens of centimeters thick.

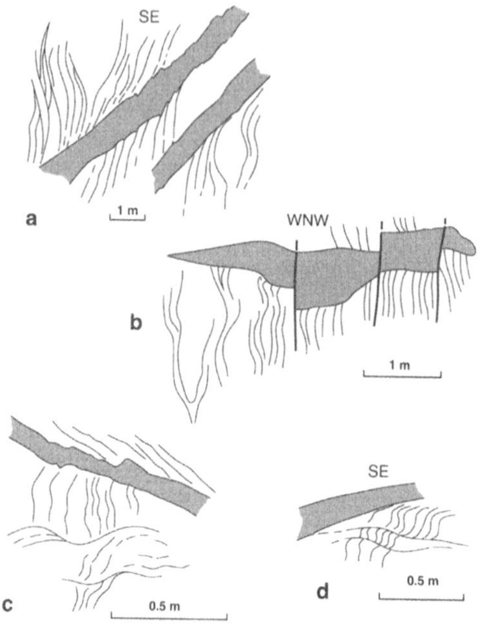

Figure 6. Relationships between basalt dikes and sheared gabbros, 4 km nortwest of Le Chenaillet, where steeply dipping foliation in gabbros with a vertical lineation is cut at high angles by dikes: (a) indented margins of thicker dike suggest that shear zones were still active during dike intrusion; (b) dike fragmented as a result of brittle deformation after magma consolidation (alternatively, dike may be affected by discrete late shears parallel to foliation plane); (c) and (d) dikes more or less parallel to late shear zones cutting protomylonitic gabbro.

Dolerites

Dolerites form up to 10 meters thick dikes and sills emplaced in the basalts. In the cores of larges masses they display a coarse-grained (up to several millimeters), sub-ophitic or intergranular texture. Such dolerites, related to a younger and now eroded volcanic pile, are unknown in the gabbroic sequence.

Structural and petrological aspects of high-temperature ductile deformation in gabbros and associated rocks

Protomylonites and anastomosing shear zones

The ubiquitous evidence, in the Montgenèvre ophiolite, for high-temperature ductile deformation affecting consolidated gabbros has already been pointed out by Mével et al. (1978) and Steen et al. (1980). The ductile deformation is lacking in ultramafic cumulates emplaced in the mantle peridotites, and progressively appears towards the top of the olivine gabbros and troctolites in the form of a faint foliation cutting the igneous layering at an angle of about 45°.

Within the overlying clinopyroxene gabbros, the deformation is highly heterogeneous and comprises an anastomosing and branched system of shear zones of highly variable thicknesses, ranging from a few milimeters to some tens of centimeters, that surround and isolate lenses and blocks of less deformed or virtually undeformed, coarse-grained gabbro (mostly Mg-gabbro) but also of amphibole-poor to amphibole-free clinopyroxene gabbro.

In many localities the deformed rocks include a brecciated facies, with angular blocks containing an incipient foliation. These angular blocks are sharply truncated at their margins by a mylonitic rim, and even by discrete schlieren of fine-grained leucodiorite. These structures suggests that deformation occurred close to the brittle-ductile transition. At the scale of individual outcrops (1 to 10 meters), the shear zones are mostly curved and may converge towards larger shear zones. Since brown amphibole is present in any deformed rock whereas it is rare or entirely absent in the adjacent undeformed gabbros, it is suggested that hydration, presumably due to access of sea water in fractures and its further percolation along shear zones and foliation planes, may have played a major role in the progressive softening and allied high-temperature recrystallization into amphibole-flaser gabbro, and possibly also in partial melting of the gabbros

Intrusive leucodiorite veins cut across the high-temperature foliation of the gabbros at angles ranging from a few degrees to nearly 90°, and occasionally display an en-echelon geometry. Young basalt dikes also cut the high-temperature foliation at variable angles, and few cases of en-echelon, lenticular and sigmoidal veins have been observed. In one case, the walls of a basalt dike show indented contacts (Fig. 6a) suggesting that the enclosing mylonitic gabbro was deforming plastically (Fig. 6b) during intrusion of the basaltic magma.

Mylonite-utramylonite belts

Along the southwestern crest of Le Chenaillet, some ultramylonite belts, up to 50 cm thick, affect all types of gabbro. The ultramylonites display a fine-scale ribbon microstructure of recrystallized plagioclase, clinopyroxene and olivine. Mylonitic deformation converted the more mafic protoliths into fine-grained amphibolites with a pronounced shape fabric of brown amphibole. Refolding of the mylonitic banding with sheath fold geometries occurs around cores of less deformed rock (Fig. 4e). The Le Soureou ultramylonites define a steep belt, 3 to 4 meters thick, with downdip, ESE to SSE trending stretching lineations. The ultramylonites cut sharply across much less deformed gabbros. The sizes and amount of clasts decrease towards the core of the mylonite belt, the matrix being progressively recrystallized and replaced and/or impregnated by fine-grained (1 mm) grainsize dioritic material (Fig. 4c).

Kinematic indicators

Good shear-sense indicators are presented by the sense of curvature of shear-induced foliations in continuous shear zones cutting isotropic gabbro (Fig. 4f). Shear senses inferred from asymmetric tails and pressure shadows tend to be inconsistent at the outcrop scale, but are coherent in mylonites and ultramylonites from planar shear zones devoid of gabbroic boudins.

Winged, rotated clasts of clinopyroxene and of altered olivine frequently display a sub-spherical shape. Diopside-salite monocrystalline clasts with straight wings of augite and/or brown amphiboles tend to show δ-type geometries (Hanmer and Paschier, 1991) presumably as a result of their mechanical strength. In contrast, σ-type geometries characterize many recrystallized mafic clasts. These geometries are developed around relics of magmatic clinopyroxenes with bent cleavages, rimmed and replaced by a fine-grained, secondary (and occasionally symplectitic) assemblage of augite and brown amphibole, and around olivine clasts replaced by serpentine and Mg-chlorite, and rimmed by tremolite-actinolite. In the rare mylonitic troctolites, relics of Ca-rich plagioclase clearly behaved totally rigid: they are sub-spherical and are free of wings and pressure shadows when enclosed in the layered ductile matrix of sepentine, chlorite, tremolite-actinolite and some ribbons of recrystallized, less calcic plagioclase. In many mafic mylonitic gabbros, plagioclase clasts may form elongate polycrystalline ribbons, less than one millimeter thick, a few millimeters wide and several centimeters in length (Fig. 4b, e).

Physical conditions and temperature constraints during the ductile deformation of gabbros

Solid-state, dry deformation is documented in few samples nearly free of amphiboles. One sample (3520) contains diopside-salite clasts of magmatic origin rimmed by fine-grained recrystallised augite (Fig. 4b and Fig. 5b) grading into polycrystalline ribbons, up to several decimeters long, progressively dispersed into less than 1 millimeter thick trails of isolated grains. This suggests that high-temperature shearing may have been initiated under dry conditions at certain horizons in the magmatic pile. Such conditions are compatible with the granulite facies, but no orthopyroxene-bearing rock has as yet been found. The widespread development of late-magmatic and syn-kinematic brown amphiboles indicates that further shearing took place under hydrous conditions, and these conditions may also have been responsible for softening of the gabbros possibly up to their melting point. On the basis of general considerations concerning the stability of ferropargasite, Mevel et al. (1978) have inferred ambient temperatures of 700-800 °C at 3-5 km depth for the intra-oceanic deformation of the gabbros. In the ferropargasite-bearing amphibolite described by Mével et al., (1978), in fact a product of mylonitization of an early mafic dioritic dike similar to that of Fig. 3e, thin veinlets slightly oblique to the metamorphic banding (Fig. 5d) clearly represent dry melts formed in situ during high-temperature deformation. These veinlets contain euhedral Fe-rich augite, apatite and zircon in a groundmass of fine-grained oligoclase. The alignment of pyroxene prisms parallel to the foliation and oblique to the vein walls point to their synkinematic crystallization. Notably, when affected a few centimeters away by further high-temperature shearing, these pyroxenes may also show a porphyroclastic microstructure rimmed or partly replaced by brown amphibole. The collection of such augite-bearing, dioritic liquids through dehydration of sheared solid amphibolites apparently requires temperatures well above 850° C (Beard, 1990; Wolf and Wyllie, 1990). It could be argued that such high temperatures may refer to cooling of the dioritic magma. However, the occurrence of augite-bearing, cross-cutting dioritic veinlets apparently formed by partial melting of amphibole-free mylonitic gabbro (Fig. 4b) also suggests temperatures equal to or exceeding 850 °C (Beard, 1990).

In samples from the Soureou ultramylonitic belt, the fine-grained (1 mm scale) matrix has been impregnated by a dioritic liquid (Fig. 4c) as evidenced by its poor planar and linear fabric defined by the alignment of sub-euhedral plagioclase, clinopyroxene and brown amphibole, the latter with poikilitic margins and including minute euhedral plagioclase.

It follows that solid-state shearing of gabbros progressively evolved into viscous flow in the core of the mylonitic belt, followed by the fine-grained crystallization of the dioritic melt and eventually ending with continued solid-state shearing. Whatever the origin of the inferred high temperatures (shear heating, or temperature increase caused by a percolating dioritic liquid), the deformation within the mylonitized gabbros and gabbro mylonites may have occurred at temperatures around 800-850° C.

Pre-Alpine brittle deformation

Several observations in the dense system of cross-cutting veinlets (some of which are filled by albite, chlorite, quartz and prehnite, and are related to reverse faulting of Tertiary age) indicate, that during the emplacement of young dolerite-basalt dikes the deformation proceeded close to the brittle-ductile transition. The occurrence of angular fragments of brecciated gabbro and of basalt in a cement of albitite are clearly related to the emplacement of young basaltic dikes. Amphibolite facies green magnesio-hornblende with significant Cl content, followed by actinolite pseudomorphs after brown amphibole along cracks may relate to brittle conditions during volcanism. Other brecciated facies with a matrix of zoisite, clinozoisite, chlorite and cross-cutting veinlets with the same minerals plus sphene and magnetite relate to younger fracturation and fluid circulation. These veinlets generally cut the foliation of the gabbros at high angles, but they also root in ductile horizons parallel to the foliation which, in such zones, is affected by intense low-temperature recrystallization with the same minerals as in the veins.

Geometry and kinematics of the high-temperature deformation

The orientations of foliation planes and shear-zones, the trends and plunges of stretching lineations and the orientations of basalt dikes have been measured, and the results are shown in Fig. 7. A first inspection suggests a considerable dispersion of stretching lineations (Fig. 7a), but this dispersion is less pronounced in smaller subareas (Fig. 7b, c, and d). This variability is due in part to later deformation associated with numerous variably oriented faults, generally with steeply oriented slickensides, that cut across the massif, and also to the effect of numerous active land slides, in particular along the Chenaillet crest. Restoration of this complex situation is apparently conjectural. However, good outcrops in steep ridges allow to draw the following conclusions, valid at the scale of some hundreds of cubic meters, which impose some kinematic constraints.

(1) Above the inferred paleo-Moho in the southwest ridge of Le Chenaillet, the incipient penetrative, high-temperature foliation trending northeast and dipping east cuts the igneous layering of the basal olivine gabbros at angles of about 45°.

(2) In the overlaying olivine-poor clinopyroxene gabbros, shear zones are curved and anastomose. Most are steep (> 75°) and trend about ESE-WNW whilst the associated stretching lineations plunge steeply to the west (pitches between 75 and 90°). At Le Soureou, gabbros exposed in the form of a linear extrusion also display steeply-dipping foliation planes, mylonite-ultramylonite zones with trends between ESE and SSE, and mostly steeply plunging stretching lineations. Kinematic indicators suggest the general downthrow of the eastern part of the massif. Shear criteria on associated low-angle shear zones are consistent with top to the WSW extensional movements and left-lateral strike-slip motion. This dispersion is primary.

(3) The major mylonite-ultramylonite zones cut across structures of already deformed gabbro and suggest continuous deformation. They may delineate mostly steeply dipping, NW-SW trending master faults.

(4) In domains of homogeneous and relatively high strains, early diorite dikes are oriented parallel to the foliation in the gabbros, and are also transformed into high-temperature protomylonites.

(5) Young basalt dikes, occasionally in conjugate sets, show evidence for

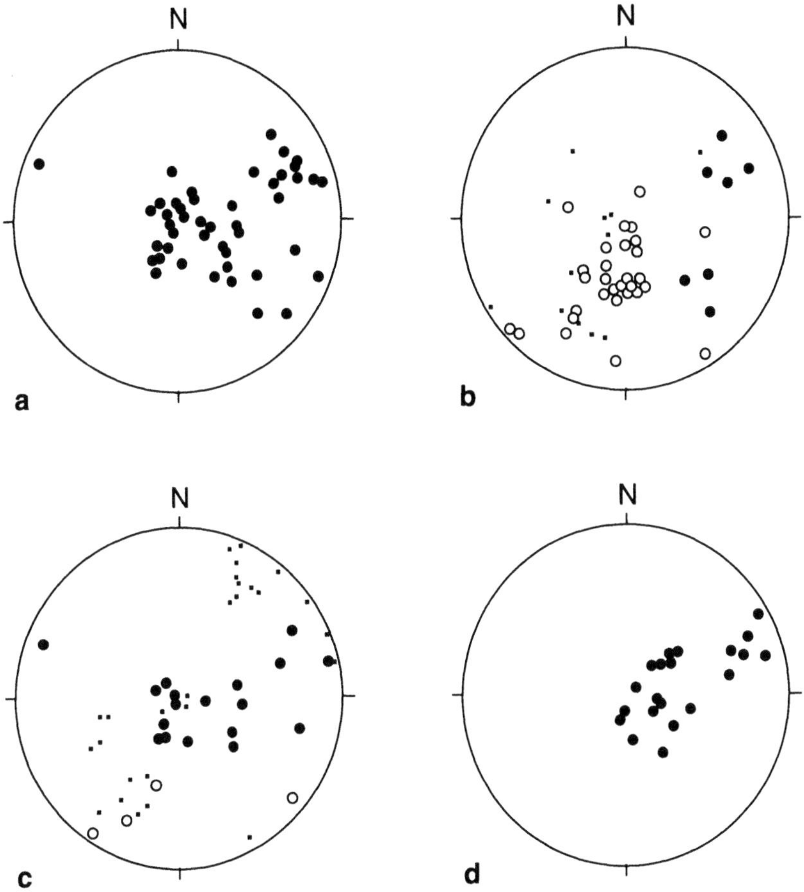

Figure 7. Equal area, lower hemisphere projections of poles to foliation planes (small dots), stretching lineations (black circles) and poles to dikes (open circles) in gabbros: (a) all areas together; (b) southwest crest of Le Chenaillet; (c) area 4 km northwest of Le Chenaillet; (d) Le Soureou-Le Lac Noir area.

rotation possibly during and certainly after their solidification. A rotation of up to 90° may be inferred for domains such as the area 4 km northwest of Le Chenaillet with a present-day horizontal geometry of dikes cutting the locally horizontal mylonitc foliation of the gabbros at high angles. The rare dikes observed in the most deformed gabbros tend to be parallel to the foliation, which may show signs of cataclastic reactivation, with some dikes being clearly boudinaged.

(6) As the younger, variably oriented extensional faults (pre-Alpine and late-Alpine) are all dominated by steeply plunging slickensides, the associated rotations must have occurred about subhorizontal axes, such that the trends of the now steeply dipping dikes (> 75°) did not seriously change and can statistically be regarded as primary. Indeed, the dikes at average

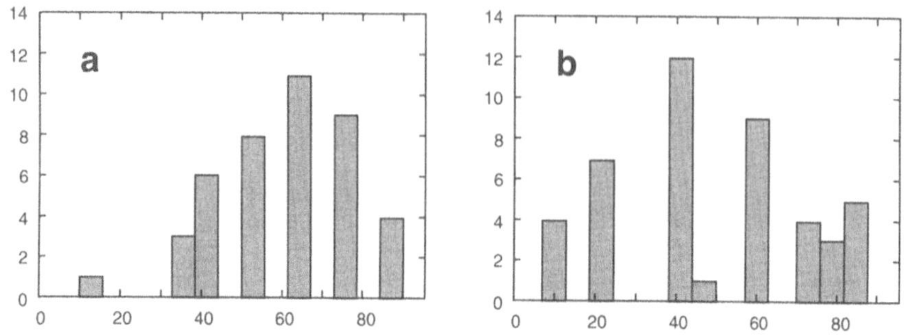

Figure 8. Histograms showing angular relationships between azimuths of basalt dikes and (a) foliation planes, and (b) stretching lineations.

do show NE trends.

(7) Although the exposure conditions, in places dominated by land slides, preclude to collect an extensive set of orientation data, the apparent main result of some tens of measurements is that the angle between azimuths of basalt dikes and the gabbro foliation planes and shear zones has a mean value bracketed between 45 and 75°, with a clear maximum around 60 to 70°, whilst the angles between azimuths of dikes and stretching lineations show two apparent maxima around 40° and 70° (Fig. 8). This is, by necessity, the only meaningful result of the orientation data measured because, for the lack of reference with respect to the paleo-horizontal or vertical, there are no further constraints on the initial geometry of the igneous layering in the gabbros, the initial dips of the basalt dikes, and the initial geometry of the shear zones. These angular relationships and the presence of en-echelon veins, sigmoidal veins, and dikes deformed prior to their complete solidification imply a genetic link between the kinematics of the deformation in the gabbros and the emplacement of the basalt dikes.

Sedimentary record of uplift and unroofing of peridotites

Arenites of gabbroic composition were directly deposited at the base of the basaltic sequence on top of the Val Gimont gabbros (Lemoine, personal communication, 1993). These detrital sediments are therefore interpreted as being related to unroofing of peridotites. Detrital sediments of mainly ultramafic nature are also exposed around Le Souréou, where they overly basalts with an assumed stratigraphic contact. The oldest formation is a non-stratified conglomerate, over 30 meters thick, made up of both rounded and angular blocks of predominantly cumulate rocks (pyroxenite, wehrlite, dunite, rare anorthosite, troctolite, chromitite) and blocks, over 1 meter diameter, of lherzolite, packed together and cemented by essentially ultramafic sands with pockets of pink micrite. These rocks are overlain by few meters of bedded, fine-grained ultramafic arenites with ultramafic, doleritic and basaltic fragments, grading into a conglomerate with predominantly rounded pebbles of dunite, a few centimeters in size and of altered basalt and diabase (Bertrand et al. 1980). This sequence is overlain, along an east-dipping contact of presumed sedimentary origin, by a basaltic breccia containing fragmented pillows and hyaloclastites which forms part of the volcanic sequence.

The Souréou ultramafic sediments, erroneously described as a "mélange" by Bertrand et al. (1980) and by Blake and Jacko (1990), in fact record the uplift of mantle peridotites synchronous with the submarine pouring of basalts. The ultramafic sediments are similar in nature, and display the same sedimentary features suggestive of mass flow deposits, as those described in Queyras at the base of the fossiliferous meta-radiolarites resting directly on top of the lherzolitic basement (Caby et al., 1987).

Discussion

Intra-oceanic plastic deformation has been observed in many ODP sites at slow-spreading ridges and transform faults. High-temperature plastic deformation affecting gabbros has been documented at Site 735 during ODP Leg 118 in the slow-spreading Southwest Indian Ridge, yielding about 30% of the rocks being plastically deformed in granulite facies to amphibolite facies conditions (Cannat et al., 1991).

Shearing at high temperature was incipient in the Montgenèvre troctolites and olivine gabbros close to the inferred paleo-Moho and involved the development of a high-temperature foliation at angles of about 45° to the igneous layering. Solid-state, high-temperature (800-850 °C) plastic deformation affected most of the overlying gabbros and produced anastomosing mylonite zones, separating lenses and blocks of less deformed to undeformed gabbro. This heterogeneity was probably not controlled by the lithology alone. The gross geometry of the deformation, with increasing strains upward in the magmatic section suggests, that it concentrated within few master faults in the rigid, olivine-bearing gabbros as well as in the underlying mantle peridotites as demonstrated by the occurrence of ultramylonitic, hydrated peridotite (Fig. 5a). It is suggested that the shear zones, possibly formed after fracture systems initiated at temperatures near the gabbro solidus, allowed the localized access of sea water and thereby enhanced the development of a heterogeneous deformation geometry.

The variable orientations of the foliation planes and shear zones at the scale of some tens to hundreds of meters may indicate considerable amounts of progressive shearing and stretching of the solidified gabbroic mass, in response to crustal extension and/or transform processes. Dioritic rocks crystallized repeatedly during ductile deformation of the gabbros. The textural relationships between the mylonitic fabrics and magmatic veins (Fig. 4a) provide strong evidence for the generation of dioritic magma by partial melting of sheared gabbro, however, the intrusion of composite dikes of microgabbro and mafic diorite in tension gashes oblique to the foliation also suggests supply of dioritic magma from depth. Most of the diorites suffered ongoing high-temperature deformation and became converted into augen gneiss and amphibolites, the most deformed ones with their foliation parallel to the foliation in the surrounding gabbros (Fig. 3e).

Young basalt dikes intruded at angles of at average 40° and 70° away from both the high-temperature foliation and the stretching lineation in the metagabbros, and there is some evidence for conjugate sets. Younger faults thoroughly affected the gabbros such that the eventual pre-Alpine deformation geometry of foliations, shear zones and dikes became rather complex. Like in several other Piemonte-Ligurian ophiolite massifs (Lemoine, 1980; Lagabrielle et al., 1984; Lemoine et al., 1987), uplift of the mantle peridotites as revealed by the deposition of ultramafic-derived sediments occurred prior to and during submarine basaltic volcanism.

Kinematic studies of deformed gabbros drilled at ODP Site 735 (Leg 118) allowed the reconstruction of normal listric faults parallel or slightly oblique to the ridge axis (Cannat et al., 1991). The final domal uplift of serpentinized peridotites in response to to continued extension has also been suggested on the basis of geometrical relationships between the different terms of a Cretaceous ophiolite observed at shallow crustal level, north of the Arabian promontory (Dilek and Delaloye, 1992). These authors also infer the existence of a low-angle master fault responsible for asymmetric extension, and consider the volcanic sequence adjacent to plutonic rocks as related to off-axis volcanism.

The extremely sinuous and anastomosing geometry of the shear zones in the Montgenèvre gabbros suggests that conjugate shear zones, like those seen at the outcrop scale, possibly formed at the scale of the oceanic crust, whereas the ultramylonitic shear zones could represent master faults. An originally steep geometry of the shear zones is preserved in the Le Soureou gabbros, which make up a linear extrusion parallel to the mylonites. The data suggest that the whole of the gabbroic layer was stretched in relation to spreading and/or transform processes during the magmatic activity. The relationships between the azimuths of the basalt dikes and the plastic deformation suggest that the dikes intruded, at decreasing temperatures, during the waning stages of the deformation in the foliated gabbros and shear zones. Since the basalt dikes also cut the mantle peridotites, they must originate from melting at much deeper levels in the mantle. This is in agreement with the distinct geochemical characteristics of the lavas as compared with those of the gabbros (Bertrand et al., 1987).

The variable azimuths of the dikes, some of which are now in an horizontal position, suggest major rotations of crustal blocks above listric faults. These rotations must in part be younger than the ductile deformation. The net effect of ductile deformation and brittle tectonics was thus an extreme thinning of the oceanic crust, which became reduced to less than 350 m of gabbros underneath less than 500 m of pillow basalts.

Conclusions

The petrological and kinematic data presented in this paper suggest that the Montgenèvre gabbros represent shallow intrusions in oceanic mantle peridotites. The layered olivine gabbros are possibly undetached from the peridotites. If so, the section of the southwest ridge of Le Chenaillet offers the first documented paleo-Moho from the western Alps.

In contrast with ophiolites derived from fast-spreading ridges such as reported in Oman (Nicolas, 1989; Ildefonse et al., 1993) very few sills of pyroxenite and dunite could be recognized at the base of the layered gabbros.

The minor generation of gabbroic magma, with all petrologic and geochemical features of oceanic crust, is consistent with a slow-spreading paleoridge (Nicolas, 1989). The gabbros rapidly cooled below 1000 °C, and strengthened while they moved away (a few kilometers presumably) from the spreading center, suffering severe plastic deformation at temperatures around 800-850 °C. Minor amounts of dioritic melt were emplaced, some of which possibly generated through partial melting of sheared gabbros. This episode occurred prior to 212±8 Ma. The solid-state deformation documented in the gabbros contrasts with the horizontal magmatic flow recorded in the much thicker pile of layered gabbros typical of fast-spreading ridges as reported from the Oman paleoridge (Nicolas, 1989; Ildefonse et al., this volume).

In the upper part of the gabbro sequence, solid-state deformation produced anastomosing high-temperature shear zones which were subsequently affected by extensive brittle deformation. An originally steep geometry is preserved within the Le Souréou-Lac Noir gabbros which represent a linear extrusion. Although the original geometry and kinematics of the shear zones in the layered gabbros could not be unambiguously restored, the orientation of the foliation, at angles of about 45° to the igneous layering in the basal olivine gabbros, suggests that these shear zones rotated at depth such as to merge into few low-angle shear zones below the Moho. As this ductile deformation predates the intrusion of the basalt dikes, it is concluded that the ductile shear zones are genetically linked to tectonic extension and transform processes during spreading and continuous denudation of rigid mantle peridotites whilst, at deeper levels, new basaltic magma was generated and extracted.

The high-temperature ductile deformation led to extreme stretching of the oceanic crust, thereby converting most of the "magmatic" Moho into a "tectonic" Moho, now defined by slices of foliated gabbro on top of mantle peridotites cut by a complex system of high-temperature extensional shear zones. The deformed layered gabbros may thus delineate a complex east-dipping accomodation zone between the plutonic sequence and the younger volcanic sequence cut by voluminous dolerite dikes. Continuous mantle denudation allowed serpentinized peridotites to become exposed in the form of protrusions on the ocean floor (Lemoine, 1980; Tricart and Lemoine, 1983; Lemoine et al., 1987), and also to form islands subject to active aerial erosion, as demonstrated by the ultramafic sediments (Caby et al, 1987). The exposed paleo-Moho became thus further converted into a "sedimentry" Moho. Like in many other Piemonte-Ligurian ophiolites, the sub-marine basaltic pile rests directly on mantle peridotites, and may result from off-axis volcanism.

Available geochronological data from the Montgenèvre ophiolite suggest that mantle denudation was roughly synchronous with the extrusion of off-axis basalts (some of which, in other Piemonte-Ligurian ophiolites, intermingle on top with radiolarites) and occurred around 165 Ma ago, ca. 50 to 60 Ma after the adjacent gabbros were accreted at the active ridge. Since spreading was accompanied by extensive transform processes (Bernoulli and Lemoine, 1980), it follows that the Piemonte-Ligurian branch of the Tethys, formed at a slow spreading rate presumably of the order of 1 to 2 cm/year, was not necessarily 500 to 1000 km wide, although such figure is not necessarily inconsistent with the overall geological data.

Acknowledgements

The author benefited greatly from discussions in the field with Catherine Mével and Marcel Lemoine, to whom this paper is dedicated. Marco Scambelluri, an anonymous reviewer and Reinoud Vissers are acknowledged for having suggested several improvements of the manuscript.

References

Beard, J. S., 1990. Dehydration melting of amphibolite at 1-7 kilobars. Trans. Am. Geophys. Union, 71: 1714.

Bernoulli, D. and Lemoine, M., 1980. Birth and early evolution of the Tethys: the overall situation. 26th Int. Geol. Congr. Paris, 1980, Coll. C5, 167-179.

Bertrand, J., Steen, D., Tinkler, C. and Vuagnat, M., 1980. The Mélange zone of the Col du Chenaillet Montgenèvre ophiolite, Hautes Alpes, France. Arch. Sci., Genève, 33: 117-138.

Bertrand, J., Courtin, B. and Vuagnat, M., 1981. Le massif ophiolitique du Montgenèvre Hautes Alpes, France, et province de Turin, Italie: Données nouvelles sur un vestige de manteau supérieur et de croûte océanique liguro-piémontais. Bull. Suisse Mineral. Pétrogr., 61: 305-322.

Bertrand, J., Courtin, B., and Vuagnat, M., 1982. Elaboration d'un secteur de lithosphère océanique liguro-piémontais d'après les données de l'ophiolite du Montgenèvre, Hautes Alpes, France, et province de Turin, Italie. Ofioliti, 7: 155-196.

Bertrand, J., Dietrich, V., Nievergelt, P., and Vuagnat, M., 1987. Comparative major and trace element geochemistry of gabbroic and volcanic rock sequences, Montgenèvre ophiolite, Western Alps. Schweiz.Mineral. Petrogr. Mitt., 67: 147-169.

Blake, M.C. and Jayko, A.S., 1990. Uplift of very high pressure rocks in the western Alps: evidence for structural attenuation along low-angle faults. In: F. Roure, P. Heitzmann and R.Polino (Editors), Deep structure of the Alps. Mem. Soc. géol. Fr., Paris, 156: 237-246.

Bodinier, J. L., 1988. Geochemistry and petrogenesis of the Lanzo peridotite body, western Alps. Tectonophysics, 149: 67-88.

Boudier, F. and Nicolas, A., 1977. Structural controls on partial melting in the Lanzo peridotites. In: H.J.B. Dick (Editor), Oregon Dept. Mineral. Ind., 96: 63-78.

Caby, R., Dupuy, C. and Dostal, J., 1987. The very beginning of the Ligurian Tethys: Petrological and geological evidence from the oldest ultramafic-derived sediments in Queyras, Western Alps, France. Eclog. geol. Helv., 80: 223-240.

Cannat, M., Mével, C. and Stakes D., 1991. Stretching of the deep crust at the slow-spreading Southwest Indian Ridge. Tectonophysics, 190: 73-94.

Carpena, J. and Caby, R., 1984. Fission-track evidence for Late Triassic oceanic crust in the French Occidental Alps. Geology, 12: 108-111.

Chapelle, B., 1990. La litosphère océanique de la Téthys ligure. Etude des magmatismes basiques et acides (massifs ophiolitiques du Montgenèvre et de Haute-Ubaye). Unpublished Thesis, Grenoble, France, 196 pp.

Dilek,Y. and Delaloye, M., 1992. Structure of the Kizildag ophiolite, a slow-spread Cretaceous ridge segment north of the Arabian promontory. Geology, 20: 19-22.

Ildefonse, B., Nicolas, A. and Boudier, F., 1993. Evidence from the Oman ophiolite for sudden stress changes during melt injection at oceanic spreading centers. Nature, 236: 673-675.

Ishiwatari, A., 1985. Alpine ophiolites: product of low-degree of melting in a transcurrent rift zone. Eart Planet. Sci. Lett. 76: 93-108.

Lagabrielle,Y., Polino, R. et al., 1984. Les témoins d'une tectonique intraocéanique dans le domaine tethysien. Analyse des rapports entre les ophiolites et leur couvertures métasédimentaires dans la zone piémontaise des Alpes franco-italiennes. Ofioliti, 9: 67-88.

Lemoine, M., 1980. Serpentinites, gabbros and ophicalcites in the Piemont-Ligurian domain of the Western Alps: possible indicators of oceanic fracture zones and associated serpentine protrusions in the Jurassic-Cretaceous Tethys. Arch. Sci. Genève, 33: 103-116.

Lemoine, M., Tricart, P. and Boillot, G., 1987. Ultramafic and gabbroic ocean floor of the Ligurian Tethys Alps, Corsica, Apennines: In search of a genetic model. Geology, 15: 622-625.

Lombado, B. and Pognante, U., 1982. Tectonic implications in the evolution of the western Alps ophiolite metagabbros. Ofioliti, 2/3: 371-394.

Mével, C., Caby, R. and Kienast, J.R., 1978. Amphibolite facies conditions in the oceanic crust: example of amphibolitized flaser-gabbro and amphibolites from the Chenaillet massif, Hautes Alpes, France. Earth Planet. Sci. Lett., 39: 98-108.

Nicolas, A., 1989. Structures of ophiolites and dynamics of oceanic lithosphere. Kluwer, Dordrecht, 367 pp.

Pupin, J. P., 1980. Zircon and granite petrology. Contr. Min. Petrol., 73: 207-220.

Hanmer, S. and Passchier, C.W., 1991. Shear sense indicators: a review. Geol. Surv. Canada Spec. Pap., 90-17: 72 pp.

Tricart, P. and Lemoine, M., 1983. Serpentite oceanic bottom in south Queyras ophiolites (French Western Alps): record of the incipient oceanic opening of the Mesozoic Ligurian Tethys. Eclogae geol. Helv., 76: 611-629.

Weissert, H.J. and Bernoulli, D., 1985. A transform margin in the Mesozoic Tethys: evidence from the Swiss Alps. Geol. Rundschau, 74: 665-679.

Wever de, P. and Caby, R., 1981. Datation de la base des schistes lustrés post-ophiolitiques par des radiolaires Oxfordien supérieur-Kimmeridgien moyen dans les Alpes cottiennes Saint Véran, France. C. Rend. Acad. Sci., 292: 467-472.

Wever De, P., Baumgartner, P. and Polino, R., 1987. Précisions sur la datation de la base des Schistes Lustrés postophiolitiques dans les Alpes Cottiennes. C. Rend. Acad. Sci., 305: 487-491.

Wolf, M.B. and Wyllie, P.J., 1990. Liquid morphology interconnectivity in solid amphibolite during dehydration melting at 10 kbar. Trans. Am.Geophys. Union, 71: 1714.

Pre-orogenic High Temperature Shear Zones in an Ophiolite Complex (Bracco Massif, Northern Apennines, Italy)

GIANCARLO MOLLI
Dipartimento di Scienze della Terra
Università di Pisa, 56126 Pisa

Abstract

This paper presents the first results of a structurally-oriented study of pre-orogenic high temperature shear zone structures which are preserved in the Northern Apennines ophiolites. A detailed analysis is made of the geometry, areal distribution and kinematics of upper amphibolite facies shear zones present in the gabbroic Bracco Massif. The high temperature shear zones show distinct deformation gradients, and three characteristic fabric types are recognized: porphyroclastic coarse grained metagabbro mylonites (fabric type 1), banded ultramylonites (fabric type 2), and "flinty" ultramylonites (fabric type 3). Microstructural observations suggest that the deformation in the gneissic mylonite types (fabric types 1 and 2) was dominated by dislocation creep, whilst a possible grainsize-sensitive flow mechanism may have been important in the very fine grained "flinty" ultramylonites (fabric type 3). Reconstruction of the shear zones into their presumed original orientations suggests that they developed as low- to medium-angle extensional fault zones. This has important implications for the inferred processes of ophiolite generation.

Introduction

In the Bracco region (Eastern Liguria) an ophiolite complex is exposed which consists of a plutonic assemblage (gabbros and ultramafic cumulates) and mantle peridotites (mainly serpentinitic lherzolites), associated with pelagic sediments representing their original sedimentary cover. This ophiolite complex is part of the Liguride Units (Elter and Pertusati, 1973), which are traditionally considered to be remnants of the Piemonte-Ligurian ocean, the western branch of the Mesozoic Tethys. Throughout the orogenic history that led to the development of the Alps and the Apennines, the Liguride Units occupied a high structural position and were only affected by very-low grade metamorphism (Pertusati and Horremberger, 1975; Hoogerduijn Strating and Van Wamel, 1989).

R.L.M. Vissers and A. Nicolas (Eds.),
Mantle and Lower Crust Exposed in Oceanic Ridges and in Ophiolites, 147-161.

Consequently, the primary features of the ophiolites are well preserved and they have been extensively studied since the end of the sixties.

In this paper, the first results are presented of a structural study which aims at unravelling the pre-orogenic tectono-metamorphic evolution of the Ligurian ophiolites. I shall focus here on some structural features of the high temperature shear zones developed in the gabbroic complex which, during the early stages of oceanic rifting, played a role in accomodating the uprise of the gabbroic and ultramafic rocks towards the ocean floor.

Geological Setting

The Northern Apennines represent a classical area for ophiolite studies. Here, Brogniart (1821) described for the first time the association of serpentinized peridotites, pillow basalts and radiolarites, later known as the 'Steinmann trilogy' (Amstuz , 1980). More recently, Decandia and Elter (1969), Bezzi and Piccardo (1971) and Galli et al. (1972) performed further research in the area. They contributed to an outline of the general features of the ophiolites and compared them with the stratigraphy of present-day oceanic lithosphere.

The general characteristics of the ophiolites are: (1) the presence of relatively undepleted mantle rocks (mainly lherzolites); (2) the absence of a true sheeted dyke complex; (3) the thin and discontinuous character of the basaltic layer; (4) the local presence of large amounts of ophiolitic breccias and (5) the direct stratigraphic juxtaposition of pelagic sediments with the ultramafic and gabbroic basement. All of these characteristics are well documented and currently accepted (Passerini, 1965; Decandia and Elter, 1972; Piccardo, 1977; Barrett and Spooner, 1977; Abbate et al., 1980; Beccaluva et al., 1980).

Another well-documented feature in the ultramafic-gabbroic complex is the petrographic evidence for a retrograde tectonic and metamorphic evolution, related to progressive denudation from initially high temperatures towards low-grade metamorphic conditions (Cortesogno et al., 1975, 1977). In the gabbros, the early part of this history which started shortly after their crystallization, is characterized by the development of high temperature shear zones. The pre-orogenic age of these shear zones is clearly indicated by cross-cutting undeformed and unmetamorphosed, younger basaltic tholeiitic dykes (Cortesogno et al., 1975).

Except for the preliminary work by Hoogerduijng Strating (1988), modern structural analysis of the deformation related to this early evolution is almost completely lacking. A precise description of the microstructural features and a study of the geometry and kinematics of the main shear zones provide a possible way to constrain interpretations of the early evolution of the Ligurian Tethys.

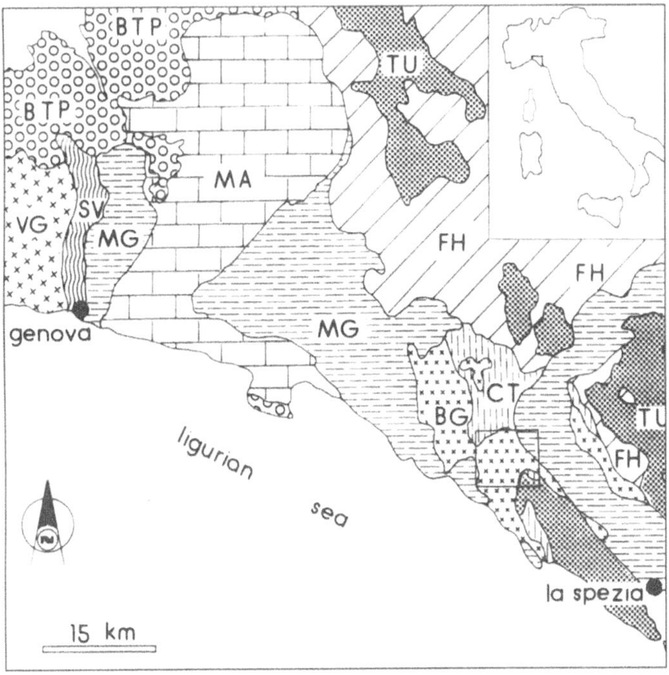

Figure 1. Tectonic sketch map of Eastern Liguria showing location of study area. BTP: Tertiary Piemontese Basin, VG: Voltri Group, SV: Sestri-Voltaggio Units, MA: Mt. Antola Unit, MG: Mt.Gottero Unit, BG: Bracco-Val Graveglia Unit, CT: Colli/Tavarone Unit, FH: External Liguride Units, TU: Tuscan and Canetolo Units (modified from Marroni, 1991).

Petrography of the host rocks and shear zones

In the studied area (Fig. 1), Mg-gabbros represent the most abundant lithotype; clinopyroxene-gabbros, olivine-gabbros and troctolites are widespread, whilst anorthosites and pyroxenites in subordinate occurrences complete the assemblage (Serri, 1980). They commonly show medium to coarse grain sizes, and gabbros with clinopyroxene crystals of up to several centimeters are frequent. A magmatic layering defined by grainsize and/or compositional variations can be observed as well as magmatic relationships between gabbros and ultramafic cumulates (plagioclase-bearing peridotites; e.g., Bezzi and Piccardo, 1970). Locally, mantle rocks crop out which, where not strongly serpentinized, can be recognized as clinopyroxene-poor, spinel lherzolites (Piccardo et al., 1992). There are primary contacts between the plutonic assemblages and the lherzolites preserved, and such contacts demonstrate the intrusive nature of the gabbro bodies into the mantle rocks. According to Cottin (1984), magmatic crystallization occurred at confining pressures around 0.5-0.7 GPa. Mg-gabbros ("Eufotidi" of early authors) consist of a magmatic paragenesis formed by euhedral or subhedral plagioclase (An 60±3) interstitial amidst poikilitic

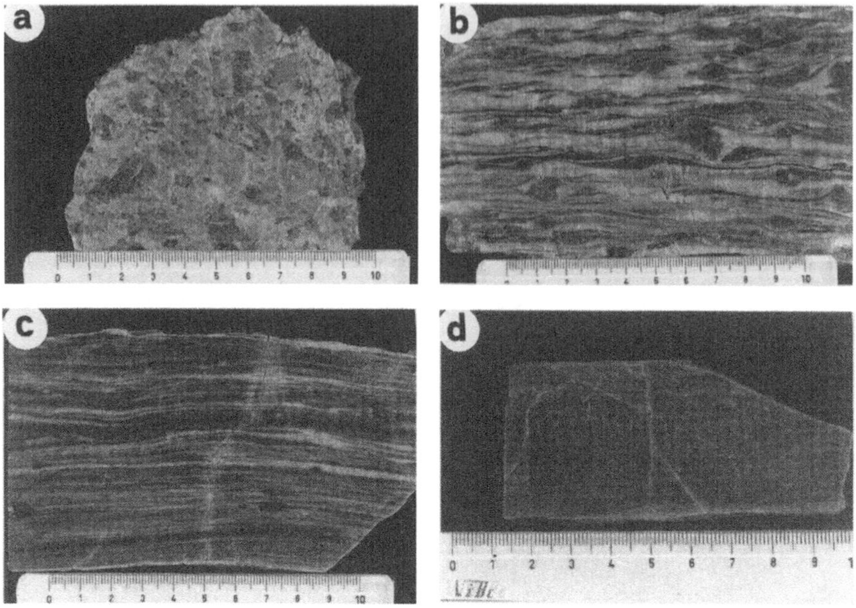

Figure 2. Polished surfaces of handspecimen illustrating principal microstructures: (a) undeformed, coarse-grained Mg-gabbro; (b) coarse-grained porphyroclastic metagabbro mylonite of fabric type 1; (c) banded metagabbro ultramylonite of fabric type 2; (d) "flinty" ultramylonite of fabric type 3.

clinopyroxene (Mg-rich diopside) and Fe-Ti oxides. Rare olivine grains are completely transformed into chlorite+actinolite aggregates. The mafic minerals usually make up less than 40% of the bulk rock volume.

The mineral assemblage in the studied shear zones is made up of Ca-plagioclase (An 55-50) + clinopyroxene (Mg-rich diopside) + scarce pargasitic hornblende (Riccardi and Tribuzio, 1992) + ilmenite, and is characteristic of upper amphibolite facies conditions (temperatures around 700 °C, pressures not exceeding 0.4 GPa; Cortesogno and Lucchetti, 1982, 1984).

Structural features of the shear zones

Within undeformed gabbros, zones of localized plastic deformation are common. These shear zones range in width from a few decimeters to several meters. The widest shear zones can be up to several tens of meters length. More common are thin zones which are only few meters long. Locally, the shear zones occur in groups, where they form anastomosing patterns of deformation zones, surrounding lenses of virtually undeformed host rocks.

In a typical shear zone, the normally coarse-grained gabbros show grain size reduction and the development of a foliation bending progressively into the high strain part of the shear zone. Locally, the shear zones are

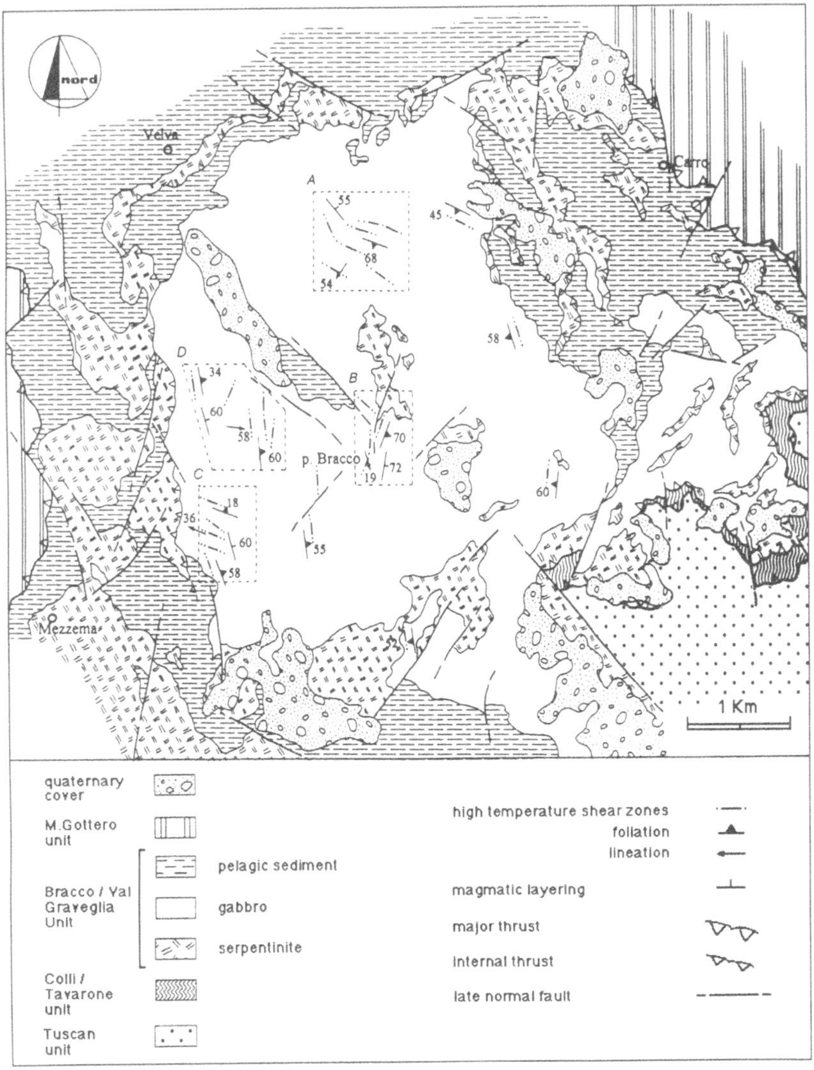

Figure 3. Schematic geological map of the Bracco area showing distribution of high temperature shear zones. Areas labelled A, B, C and D have major concentrations of shear zones.

characterized by abrupt changes in microstructure along the edges of high-strain domains, similar to the "discontinuous" or "discrete" shear zones described by Burg and Laurent (1978) and Vauchez (1987).

Within the undeformed isotropic Mg-gabbros, three types of shear zone fabrics can be distinguished (Fig. 2). These three types, representative of a progressive strain gradient, are: porphyroclastic coarse grained metagabbro mylonite (fabric type 1), banded ultramylonite (fabric type 2), and "flinty" ultramylonite (fabric type 3). The first and the second type are

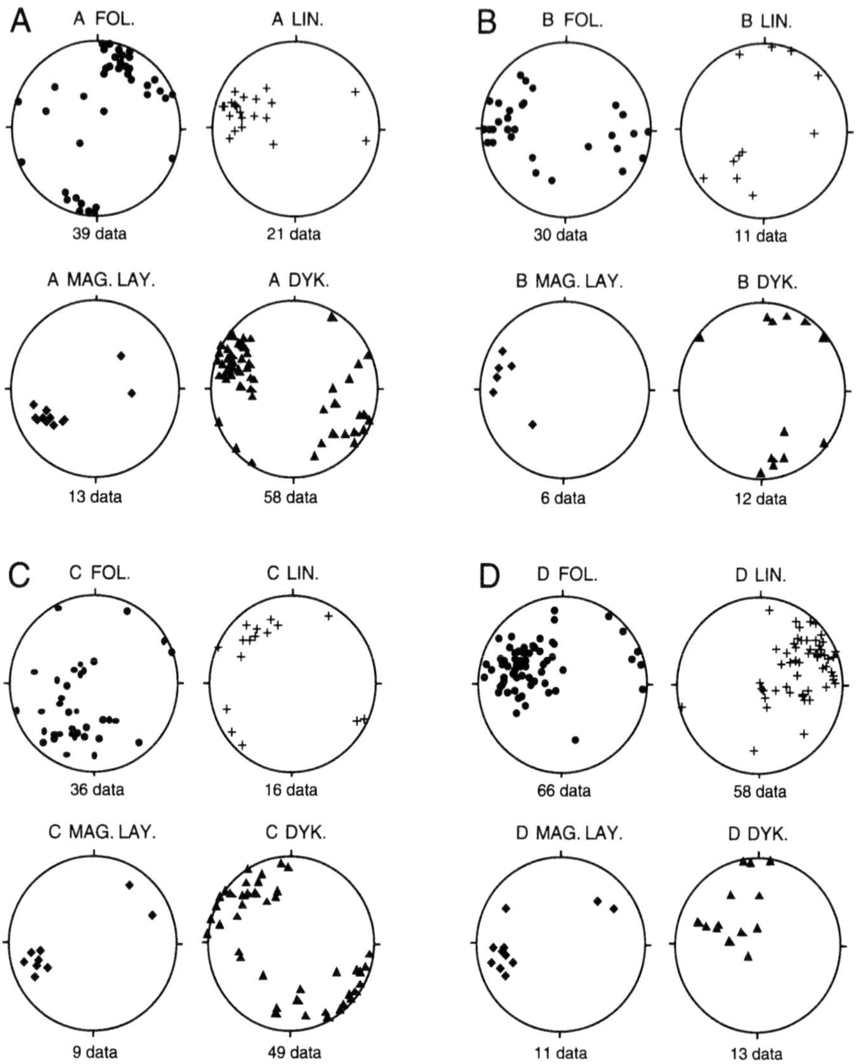

Figure 4. Compilation of structural data from areas indicated in Fig. 3. All plots equal area, lower hemisphere projections. Black dots: poles to shear foliation (FOL); crosses: stretching lineations (LIN); black diamonds, poles to magmatic layering (MAG LAY); triangles, poles to basaltic dykes (DYK).

very common; the "flinty" ultramylonites are developed at a mesoscopic scale in larger shear zones only. At the microscopic scale they also occur as millimeter-scale bands in between the other two fabric types.

Detailed geological mapping (Fig. 3) has been carried out to assess the geometry and kinematics of the shear zones. At each location, the attitude of the shear zones, magmatic layering and basaltic dykes were measured (Fig. 4). Kinematic analysis has been carried out using mineral lineations to determine the shear direction, and kinematic indicators to assess the sense of movement. At the mesoscopic scale the most reliable indicator is the deflection, observed on suitable surfaces, of shear-induced foliations into the high strain parts of the shear zones. At the microscopic scale, the criteria used to assess the movement sense include:

- the asymmetric geometry of pyroxene porphyroclasts such as σ-type (Passchier and Simpson, 1986) or "book-shelf " types (Fig. 5a,b);
- asymmetric folds, deforming the fine-grained pyroxene-rich bands which are frequently found in association with porphyroclasts (Fig. 5c);
- shape preferred orientations of plagioclase oblique to the main foliation (Lister and Snoke, 1984; Fig. 5d) which turned out to be the most useful kinematic indicator within fabric type 2 (banded ultramylonite);
- shear bands or extensional crenulation cleavages (Platt and Vissers, 1980) formed at angles of 10-20° to the main foliation. These are sometimes found in the high strain parts of the shear zones and are in particular associated with fabric type 3.

Since Mg-gabbros are the most common lithotype in the area, we concentrate below on a description of the three abovementioned mylonitic fabric types developed in these rocks. Field and microstructural relationships show that there are gradual variations within each of the fabrics and suggest that these variations could represent progressive stages in the deformational history.

Fabric type 1 (Fig. 2b) includes porphyroclastic coarse-grained metagabbro mylonites (also referred to as "flaser gabbro") which mainly consist of "augen" of clinopyroxene in a fine-grained plagioclase matrix. The "augen" are composed of cores of magmatic pyroxene (mm to cm size) surrounded by polygonal pyroxene neoblasts (grain sizes 0.2-0.1 mm). The neoblasts also make up recrystallized tails.

The recrystallization of the plagioclase initially occurs along the grain boundaries and a progressive grain size refinement can be observed. The typical grain size in this fabric type is about 0.1 mm. The grain contacts of the recrystallized plagioclase show straight or curved boundaries.

Ultrafine grained clinopyroxene and plagioclase in millimetric bands appear near transitions to fabric type 2. In these bands, the mineral grain sizes are strongly reduced to less than 50 μm.

Fabric type 2 (Fig. 2c) are banded metagabbro ultramylonites, characterized by a millimeter-scale alternation of pyroxene-rich (around 0.1 mm in size) and plagioclase-rich (0.05 - 0.08 mm) layers. On a mesoscopic scale, this microstructure has a distinct banded appearence. Magmatic clinopyroxenes are almost completely absent in rocks showing fabric type 2, and Ti-rich hornblende, present as an interstitial phase between the clinopyroxene grains, is interpreted as a late phase with respect to the deformation.

The recystallized plagioclase shows various types of grain contacts. Both

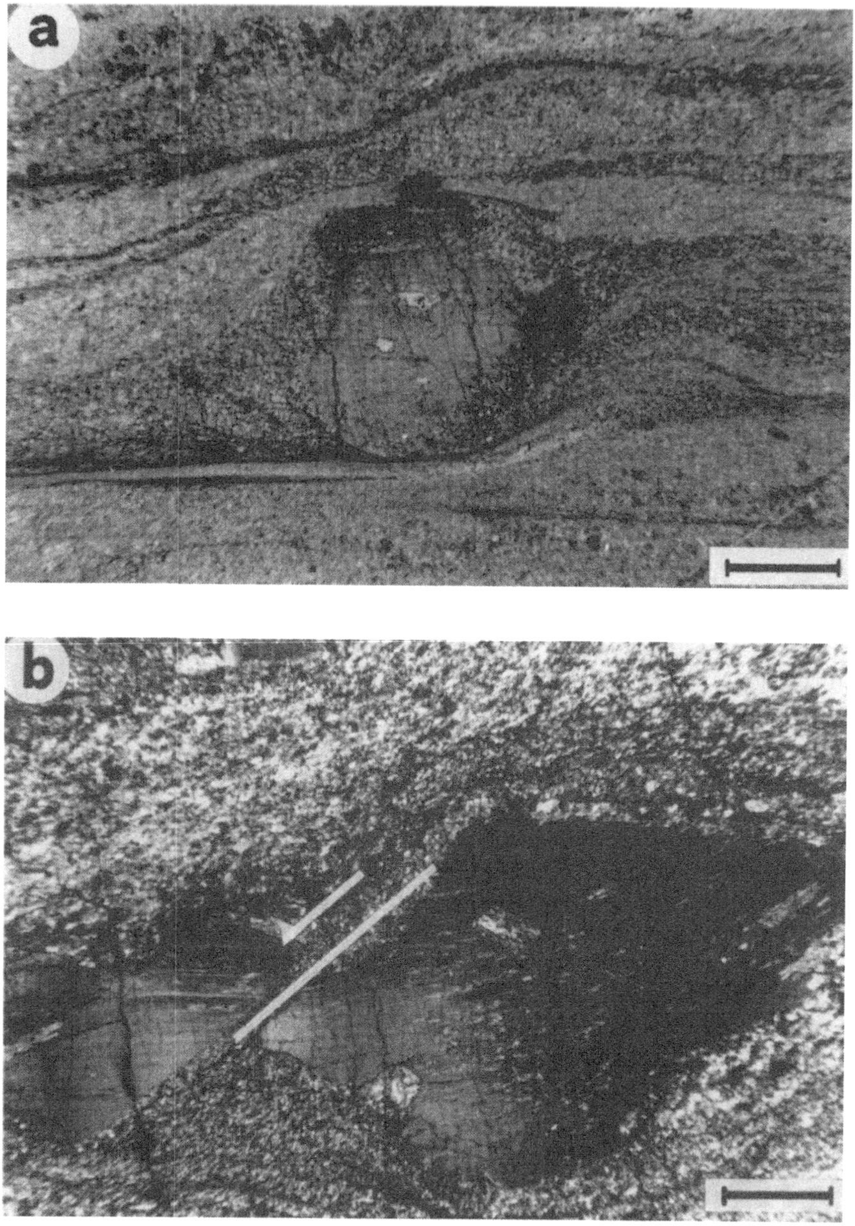

Figure 5. Microstructures illustrating kinematic indicators used, in sections perpendicular to foliation and parallel to stretching fabric: (a) σ-type clinopyroxene porphyroclast, plane polarized light, scale bar 2 mm; (b) book-shelf type clinopyroxene porphyroclast, with antithetic microfault, crossed nicols, scale bar 2 mm.

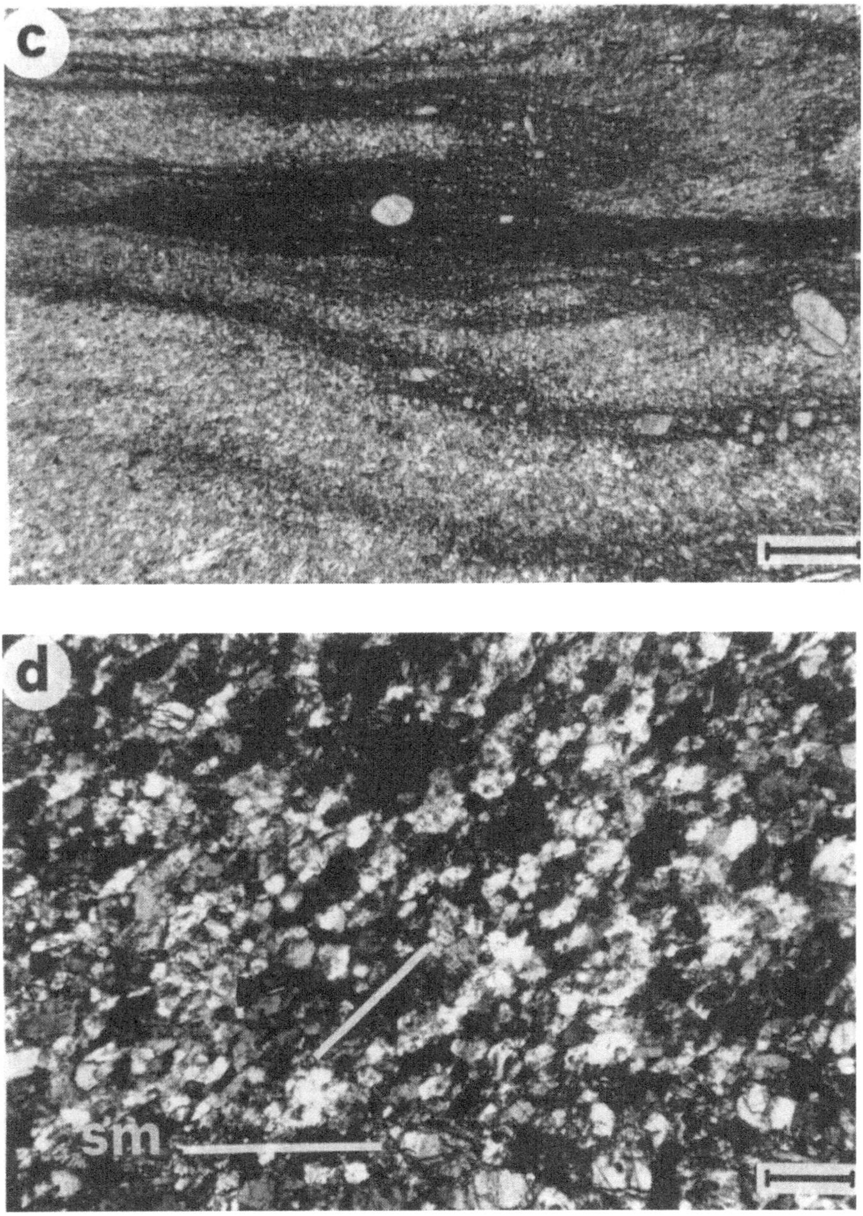

Figure 5 continued. (c) asymmetric isoclinal microfold, crossed nicols, scale bar 1 mm. (d) plagioclase grain-shape fabric oblique to mylonitic foliation (Sm) in banded metagabbro ultramylonite of fabric type 2, crossed nicols, scale bar 0.25 mm.

straight and curved to strongly irregular boundaries can be observed within the same thin section. These different contacts are interpreted as representing different degrees of recovery associated with grain growth. Locally, plagioclase shows a well-developed microstructure of irregular grain boundaries and a strong shape preferred orientation oblique to the main foliation. Ultrafine-grained, sub-millimeter scale bands of clinopyroxene and plagioclase are more frequent than in fabric type 1.

Fabric type 3 (Fig. 2d) comprises "flinty" ultramylonites, characterized by extreme grain size reduction. Both the plagioclase and the pyroxene have grain sizes less than 50 μm. A sub-millimeter scale banding only observable in thin section is formed by pyroxene-rich bands alternating with pyroxene and plagioclase-rich bands. Both minerals show a predominantly equiaxed shape fabric, although the pyroxenes sometimes show a grain shape preferred orientation parallel to the main foliation. Isolated larger pyroxenes (0.1-0.2 mm in size) are present in the ultrafine grained pyroxene-rich bands, and they may be relics derived from recrystallized pyroxene layers.

Microstructural features of fabric types 1 and 2 suggest crystal-plasticity associated with dynamic recrystallization as the predominant deformation mechanism in both the plagioclase and clinopyroxene grains. A possible change from predominantly dislocation creep to a grain size sensitive flow within the ultrafine grained pyroxene and plagioclase-rich bands (fabric type 3) is suggested by the very fine grain sizes of the phases (less than 50 μm), their extensive mixing and their predominantly equiaxed grain shapes.

Geometric and kinematic analysis

The shear zones occur throughout the Bracco region, however, four different areas (see inset in Fig. 3) have been identified in which they are ubiquitous and show a rather uniform orientation. Within these areas, a general relationship is observed of the shear zones being oriented at low to medium angles to the magmatic layering.

The reconstruction of the pre-orogenic deformation geometry in ophiolitic complexes represents a crucial point for a better definition of the original paleotectonic setting (Nicolas, 1989). According to Nicolas and Violette (1982), an interpretative framework can be reconstructed using:

a) planes of deposition in the volcanic and overlaying sediments,
b) attitudes of basaltic dykes in sheeted dyke complexes,
c) major lithologic contacts within mafic and ultramafic units,
d) internal structures in the gabbros, e.g., magmatic layering, lamination and possibly magmatic lineations,
e) compositional banding and deformation structures in mantle tectonites.

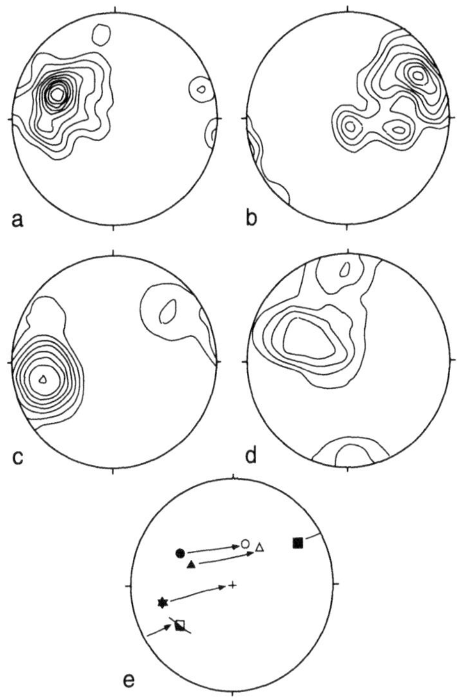

Figure 6. Reconstruction of initial deformation geometry in area D: (a) 66 poles to mylonitic foliation; (b) 58 stretching lineations on high temperature shear foliations; (c) 11 poles to magmatic layering; (d) 13 poles to basaltic dykes. All plots (a through d) equal area, lower hemisphere projections, contoured at 1, 2, 3, etc. times uniform; (e) geometric restoration on the basis of inferred horizontal orientation of magmatic layering (black star). Average orientations (concentration maxima) of other structural features are rotated over same angle about same rotation axis (circle: average pole to foliation, square: average lineation, triangle: average pole to dykes, filled symbols unrotated orientations, open symbols after rotation). Split square represents restored average orientation of lineation, with black side pointing to downthrown block.

Due to the characteristic features of the Ligurian ophiolites exposed in the Bracco area we can only use the contact of the sediments, in this case represented by Palombini shales (e.g., Hoogerduijng Strating, 1988), and the internal structures in the cumulates. In some areas (e.g., area A in Fig. 3) the magmatic layering in the gabbros and in the ultramafic cumulates appears roughly parallel (angles less than 10°) to each other and to the contact between these two lithotypes. Therefore, the magmatic layering in the gabbros has been inferred to represent a paleohorizontal surface, also in areas where no ultramafic cumulates are present. This is in agreement with the tenet that in the middle-lower part of a magmatic chamber the layering is close to horizontal (Casey et al., 1983; Nicolas, 1989), an interpretation already accepted for the Bracco cumulates by other workers (e.g., Cortesogno et al., 1987) .

In order to reconstruct the deformational geometry in the four areas relative to a paleohorizontal reference frame, the average orientation of the magmatic layering is assumed to have been horizontal, and all shear zones and their kinematic indicators are rotated accordingly. The result, illustrated in Fig. 6e, suggests a low to medium angle of the shear zones with respect to the paleohorizontal, a predominant down-dip orientation of the lineations, and a general extensional character of the fault zones.

The present result is largely consistent with the preliminary conclusions achieved by Hoogerduijn Strating (1988) in the same area, notably with regard to the kinematics of the shear zones. This is particularly significant because the datum plane used in the two studies was different: Hoogerduijn Strating (1988) used the pelagic sediment/basement contact as a reference surface, whereas in this study I have used the magmatic layering and ultramafic cumulate/gabbros relationships.

Discussion

The overall interpretation of the tectonic environment in which the Northern Apennine Ophiolites developed is still subject to debate. Various authors in the last decades have proposed a transform setting (Gianelli and Principi, 1974; Abbate et al., 1980; Cortesogno et al., 1987, and references therein). According to this model, the deformation in the gabbroic rocks is due to transcurrent movements and to diapiric emplacement of upper mantle rocks into a transform zone. However, the structural data from Hoogerduijn Strating (1988) and those presented here are inconsistent with a transform fault setting, since the latter would require the presence of more pervasive zones of deformation associated with a predominantly lateral sense of movement (Choukroune et al., 1978).

Evidence for plastic deformation is not only found in the plutonic complex but also in the peridotites. Unfortunately, however, extensive serpentinization hampers an analysis of these structures. Currently available observations allow to suggest a complex polyphase deformation history, characterized by an early tectonite fabric with spinel (?) bearing assemblages, developed in domains of several tens of meters width, followed by the local development of peridotitic mylonites showing recrystallization of olivine + clinopyroxene + brown hornblende + plagioclase. The peridotitic mylonites may be correlated with the high temperature shearing seen in the gabbroic rocks (Cortesogno et al., 1979). Subsequent serpentinite mylonites, with ribbon microstructures and recrystallization of serpentine + tremolite + magnetite + chlorite developed during ongoing deformation under decreasing temperature conditions.

A very similar tectonic evolution has recently been described by Drury et al. (1990) and Vissers et al. (1991) from the Erro-Tobbio lherzolites of the Voltri Group (see Fig. 2 for location). Consistent with previous studies by Decandia and Elter (1969), Piccardo (1977), Lombardo and Pognante (1982) and Lemoine et al. (1987), their structural studies strongly support the idea that the initial stages of opening of the Piemonte-Ligurian ocean were accomodated by the development of shear zones eventually leading to tectonic denudation of the upper mantle. As a working hypothesis, I suggest that the deformation of the gabbros in the Bracco area can be understood in the context of this general model, the shear zones being developed such as to accomodate for strain incompatibilities during mantle upwelling.

Conclusions

In this paper, data pertaining to the microstructure, geometry and kinematics of high temperature shear zones of the gabbroic Bracco massif have been presented. The different fabrics developed in the metagabbro mylonites can be related to a progressive increase and localization of the strain, and to a possible change in predominant deformation mechanism. A restoration of the deformation geometry relative to an interpreted paleohorizontal reference frame points to a low to medium angle between the shear zones and the inferred paleohorizon, and to a predominantly extensional character of the shear zones. The studied deformation of the gabbroic rocks can be interpreted in the context of a general model of rifting, breakup and development of the oceanic Ligurian Tethys accomodated by localized deformation in shear zones.

Acknowledgements

I should like to thank Kate Brodie and Ernie Rutter for their helpful suggestions and discussion about the shear zones, and for their kind hospitality during a visiting research period at the Department of Geology, Manchester University. Eilard Hoogerduijng Strating and an anonymous reviewer are thanked for careful reviews and improvement of the text.
P. Elter, G. Serri, L. Cortesogno, G.B. Piccardo and R.L.M. Vissers are thanked for discussion on the northern Apennine ophiolites. Structural data were processed using program Stereoplot II by Neil Mancktelow. This study was supported by MURST funds (P. Elter) and by the Centro Studi Geologia Strutturale e Dinamica Appennino Settentrionale (CNR, dir. A. Rau).

References

Abbate, E., Bortolotti, V. and Principi, G., 1980. Apennine ophiolites: a peculiar oceanic crust. In: Rocci G. (Editor), Tethyan Ophiolite Special issue, Ofioliti, 1: 59-96.

Amstuz, G.C., 1980. The early history of the term ophiolites and its evolution until 1945. In: Geol. Surv. Dept. (Editor) Proceedings Intern. Ophiolite Symposium, Cyprus, 149-152.

Barrett, T.J. and Spooner, E.T.C., 1977. Ophiolitic breccias associated with allochthonous oceanic crustal rocks in East Ligurian Apennines, Italy - A comparison with observations from rifted oceanic ridges. Earth Planet. Sci. Lett., 35: 79-91.

Beccaluva, L., Piccardo, G.B. and Serri, G., 1980. Petrology of the northern Apennine ophiolites and comparison with other Tethyan ophiolites. In: Geol. Surv. Dept. (Editor) Proceedings Intern. Ophiolite Symposium, Cyprus, 314-330.

Bezzi, A. and Piccardo, G.B., 1970. Studi petrografici sulle formazioni ofiolitiche della Liguria. Riflessioni sulla genesi dei complessi ofiolitici in ambiente appenninico ed alpino. Rend. Soc. It. Min. Petr., 26: 1-42.

Bezzi, A. and Piccardo, G.B., 1971. Structural features of the Ligurian ophiolites: petrological evidence for the "oceanic" floor of the Northern Apennine geosyncline; a contribution to the problem of the alpine-type gabbro-peridotite associations. Mem. Soc. Geol. It., 10: 53-63.
Brogniart, A., 1821. Sur le gisement ou position relative des ophiolites, euphotides, jaspes etc. dans quelques parties des Apennins. Ann. des Mines, 6: 177-238.
Burg, J.P. and Laurent, Ph., 1978. Strain analysis of a shear zone in a granodiorite. Tectonophysics, 39: 121-139.
Casey, J.F., Karson, J.A., Elthon, D., Rosencrantz, E. and Titus, M., 1983. Reconstruction of the geometry of accretion during formation of the Bay of Islands ophiolite complex. Tectonics, 2: 509-528.
Choukroune, P., Francheteau, J. and Le Pichon, X., 1978. In situ observation along transform fault "a" in the Famous area: Mid Atlantic ridge. Geol. Soc. Am. Bull., 89: 1013-1029.
Cortesogno, L., Gianelli, G. and Piccardo, G.B., 1975. Pre-orogenic metamorphic and tectonic evolution of the ophiolite mafic rocks (Northern Apennine and Tuscany). Bull. Soc. Geol. It., 94: 291-321.
Cortesogno, L., Gianelli, G. and Messiga, B., 1977. Le rocce gabbriche dell'Appennino settentrionale: III. Evoluzione metamorfica in ambiente oceanico e orogenico, confronto con metagabbri a metamorfismo alpinotipo. Ofioliti, 2: 75-114.
Cortesogno, L., Galbiati, B. and Principi, G., 1979. Le brecce serpentinitiche giurassiche della Liguria orientale. Arch. Sci. Gèneve, 33: 185-200.
Cortesogno, L. and Lucchetti, G., 1982. Il metamorfismo oceanico nei gabbri ofiolitici dell'Appennino Ligure: aspetti mineralogici e paragenetici. Rend. Soc. It. Min. Pet., 38: 561-579.
Cortesogno, L. and Lucchetti, G., 1984. Ocean floor metamorphism of metagabbros and striped amphibolites (T.Murlo, Southern Tuscany). Neues Jahrbuch Miner. Abh., 148: 276-300.
Cortesogno, L., Galbiati, B. and Principi, G., 1987. Note alla "carta geologica delle ofioliti del Bracco" e ricostruzione della paleogeografia giurassico-cretacea. Ofioliti, 12: 261-342.
Cottin, J.Y., 1984. Les gabbros filoniennes recoupant les lherzolites à spinelle et plagioclase du Bracco (Apennins-Ligures, Italie). Bull. Soc. Géol. France, 26: 935-944.
Decandia, A. and Elter, P., 1969. Riflessioni sul problema delle ofioliti nell'Appennino settentrionale (Nota preliminare). Atti Soc. Tosc. Sci. Nat., 75: 1-9.
Decandia, A. and Elter, P., 1972. La zona ofiolitifera del Bracco nel settore compreso tra Levanto e la Val Graveglia (Appennino Ligure). Mem. Soc. Geol. It., 11: 503-530.
Drury, M.R., Hoogerduijn Strating, E.H. and Vissers, R.L.M., 1990. Shear zone structures and microstructures in mantle peridotites from the Voltri massif, Ligurian Alps, N.W. Italy. Geologie en Mijnbouw, 69: 3-17.
Elter, P., and Pertusati, P.C., 1973. Considerazioni sul limite Alpi-Apennino e sulle relazioni con l'arco delle Alpi Occidentali. Mem. Soc. Geol. It., 12: 359-375.
Galli, M., Bezzi, A., Piccardo, G.B., Cortesogno, L. and Pedemonte, G.M., 1972. Le ofioliti dell'Appennino Ligure: un frammento di crosta-mantello "oceanici" dell'antica Tetide. Mem. Soc. Geol. It., 11: 467-502.
Gianelli, G. and Principi, G., 1974. Studies on mafic and ultramafic rocks: Breccias of the ophiolitic suite in the M. Bocco area (Ligurian Apennine). Boll. Soc. Geol. It., 93: 277-308.
Hoogerduijn Strating, E.H., 1988. High temperature shear zones in the gabbroic Bracco massif (N.Apennines, Italy): possible implications for tectonic models of ocean floor generation. Ofioliti, 13: 111-126.
Hoogerduijn Strating, E.H and Van Wamel, W.A., 1989. The structure of the Bracco Ophiolite complex (Ligurian Apennines, Italy): a change from Alpine to Apennine polarity. J. Geol. Soc. London, 146: 933-944.
Lemoine, M., Tricart, P. and Boillot, G., 1987. Ultramafic and gabbroic ocean floor of the Ligurian Tethys (Alps, Corsica, Apennines): In search of a genetic model. Geology, 15: 622-625.
Lombardo, B. and Pognante, U., 1982. Tectonic implications in the evolution of the Western Alps ophiolites metagabbros. Ofioliti, 7: 371-394.
Lister, G.S. and Snoke, A.W., 1984. S-C Mylonites. J. Struct. Geol., 6: 617-638.
Nicolas, A., 1989. Structures of ophiolites and dynamics of oceanic lithosphere. Kluwer Acad. Publ., 367 pp.
Nicolas, A. and Violette, J.F., 1982. Mantle flow at oceanic spreading centers: models derived from ophiolites. Tectonophysics, 81: 319-339.
Passchier, C.W. and Simpson, C., 1986. Porphyroclast systems as kinematic indicators. J. Struct. Geol., 8: 831-843.
Passerini, P., 1965. Rapporti tra le ofioliti e le formazioni sedimentarie tra Piacenza ed il Mar Tirreno. Boll. Soc. Geol. It., 84: 93-176.

Pertusati, P.C. and Horrenberger, J.C., 1975. Studio strutturale degli scisti di Val Lavagna (Unità del Monte Gottero, Apennino Ligure). Bol. Soc. Geol. It., 94: 1375-1436.

Piccardo, G.B., 1977. Le ofioliti dell'areale ligure: petrologia e ambiente geodinamico di formazione. Rend. Soc. It. Min. Pet., 33, 221-252.

Piccardo, G.B., Rampone, E. and Vannucci, R., 1992. Ligurian peridotites and ophiolites: from rift to ocean floor formation in the Jurassic Ligure-Piemontese basin. Acta Vulcanologica, Marinelli Volume, 2: 313-325.

Platt, J.P. and Vissers, R.L.M., 1980. Extensional structures in anisotropic rocks. J. Struct. Geol., 2: 397-410.

Riccardi, M.P. and Tribuzio, R., 1992. Gli anfiboli nei gabbri dell'Appennino Settentrionale: aspetti composizionali e cristallochimici. In: Riassunti del 76° Congresso Società Geologica Italiana (Firenze, 21-23 Sett. 1992): p.82.

Serri, G., 1980. Chemistry and Petrology of gabbroic complex of the northern Apennine ophiolites. In: Geol. Surv. Dept (Editor), Proceedings Intern. Ophiolite Symposium, Cyprus, 296-313.

Vauchez, A., 1987. The development of discrete shear-zones in a granite: stress, strain and changes in deformation mechanisms. Tectonophysics, 133: 137-156.

Vissers, R.L.M., Drury, M.R., Hoogerduijn Strating, E.H. and Van der Wal, D., 1991. Shear zones in the upper mantle: A case study in an Alpine lherzolite massif. Geology, 19: 990-993.

A Detailed Study of Mantle Flow away from Diapirs in the Oman Ophiolite

B. ILDEFONSE, S. BILLIAU and A. NICOLAS
Laboratoire de Tectonophysique, Université Montpellier II - CNRS,
Place Bataillon, 34095 Montpellier Cedex 05, France.

Abstract

Shear direction inversions in uppermost mantle flow are documented in several ophiolites. We relate such inversions to the forced flow away from the top of mantle diapirs. In the Oman ophiolite, however, the shear direction inversion is not systematically developed at the same level in the ophiolite sequence. In particular, in the vicinity of mantle diapirs, the uppermost mantle forced flow diverging from the diapir has a clear outward (top away from the diapir) direction of shear, whilst the inward (top toward the diapir) relative shear direction, due to the forced flow, seems to occur within the thick crystal mush of the overlying magma chamber. Conversely, away from diapirs, the inward shear direction is recorded in the top mantle section. A detailed study conducted in the peridotite section of the Hilti massif, presumably located at some distance to the South-East of its parent diapir, reveals that the shear direction inversion occurred at a variable depth with respect to the Moho. The observed structure suggests the existence of mantle flow channels just below the Moho, with comparatively more stagnant zones in between.

Introduction

Progress in the understanding of rock plasticity in terms of physical conditions of the deformation and kinematics of plastic flow (Nicolas and Poirier, 1976) has resulted in the possibility of analyzing peridotite structures of ophiolites in terms of asthenospheric or lithospheric mantle flow. In ophiolite massifs, where a paleo-ridge reference can at least partially be reconstructed, systematic mapping of peridotite structures combined with studies on microstructures and crystallographic fabrics allow us to trace the asthenospheric flow pattern below the ridge. This is how mantle diapirs were first discovered and mapped in a few ophiolite massifs (Nicolas and Violette, 1982). Simultaneously it was discovered that, in various ophiolite areas where the mantle flow just below the Moho was horizontal, the high

R.L.M. Vissers and A. Nicolas (Eds.),
Mantle and Lower Crust Exposed in Oceanic Ridges and in Ophiolites, 163-177.

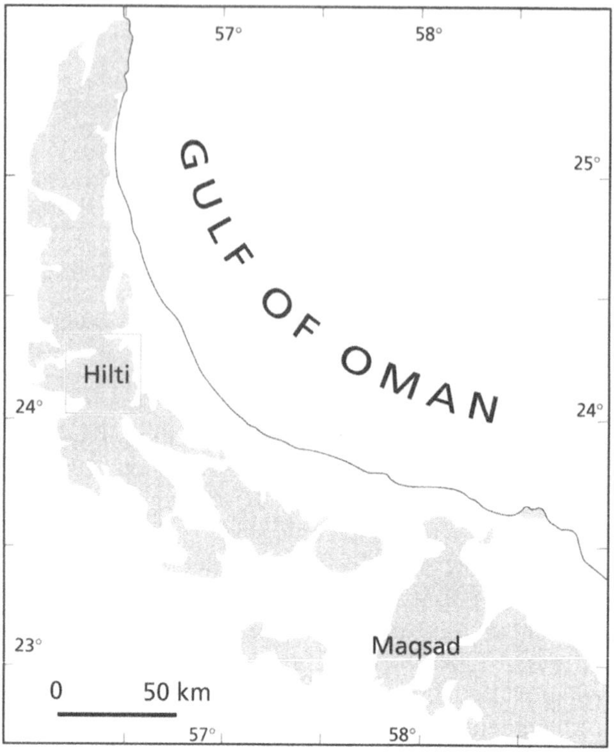

Figure 1. Location of the Hilti and Maqsad massifs in the Oman ophiolite.

temperature flow presented a shear direction inversion at a few hundred meters to a few kilometres below the Moho (Cassard, 1980; Prinzhofer et al., 1980; Violette, 1980; Girardeau and Nicolas, 1981; Podvin, 1983; Suhr, 1992; Hoxha, 1993). It has been suggested that this shear inversion was induced by a forced flow, due to mantle material being expelled from a nearby diapir, and this interpretation has inspired an active mantle diapirism model (Rabinowicz et al., 1984 ; Rabinowicz et al., 1987).

More extensive field studies in the Oman ophiolites (Fig. 1) have not systematically shown this shear inversion below the Moho (Nicolas and Boudier, in press). In the vicinity of diapirs, at any depth attained in the ophiolite (5-10 km) below the Moho, the shear is in most cases top-away from the diapir, without any shear inversion. However, the shear inversion may be locally observed very close to the Moho (Fig. 2; Ceuleneer, 1991). In contrast, in areas away from any visible diapir such as in the Hilti massif in the northern part of the ophiolite belt (Fig. 1), Ceuleneer et al. (1988) have documented a shear direction inversion, but in this earlier study the inversion appeared less systematic than in other ophiolites.

Below we report a detailed structural and kinematic study conducted in the Hilti massif aiming to map the shear direction pattern over a large area,

Figure 2. Structural map of the diapiric area in the Maqsad massif (location shown on Figure 1). Top-to-the-W and top-to-the-E movement directions indicated by black and white arrows, respectively. The lineation isodip (45°) defines a closed area where the lineation plunges more than 45°. The trend of the ridge axis is given by the trend of the sheeted dike complex in the Maqsad massif. In the northern part of the massif, the outward shear flow is recorded on both sides of the ridge axis; in the southern part, the partitioning is less clear and this may reflect a shear inversion in the uppermost mantle very close to the Moho (dominant inward shear senses at the eastern side of the diapir). See Nicolas and Boudier (in press) for further discussion.

and to investigate how this domain relates to domains closer to diapirs where inversion of the shear direction is not observed. The choice of this massif is also justified by its remarkably regular flow pattern in the mantle section below a flat-lying and planar Moho. Ultimately, such detailed structural information on the 3D geometry of large-scale mantle flow near a paleo-ridge may help to improve our knowledge on mantle diapirism and related ridge dynamics.

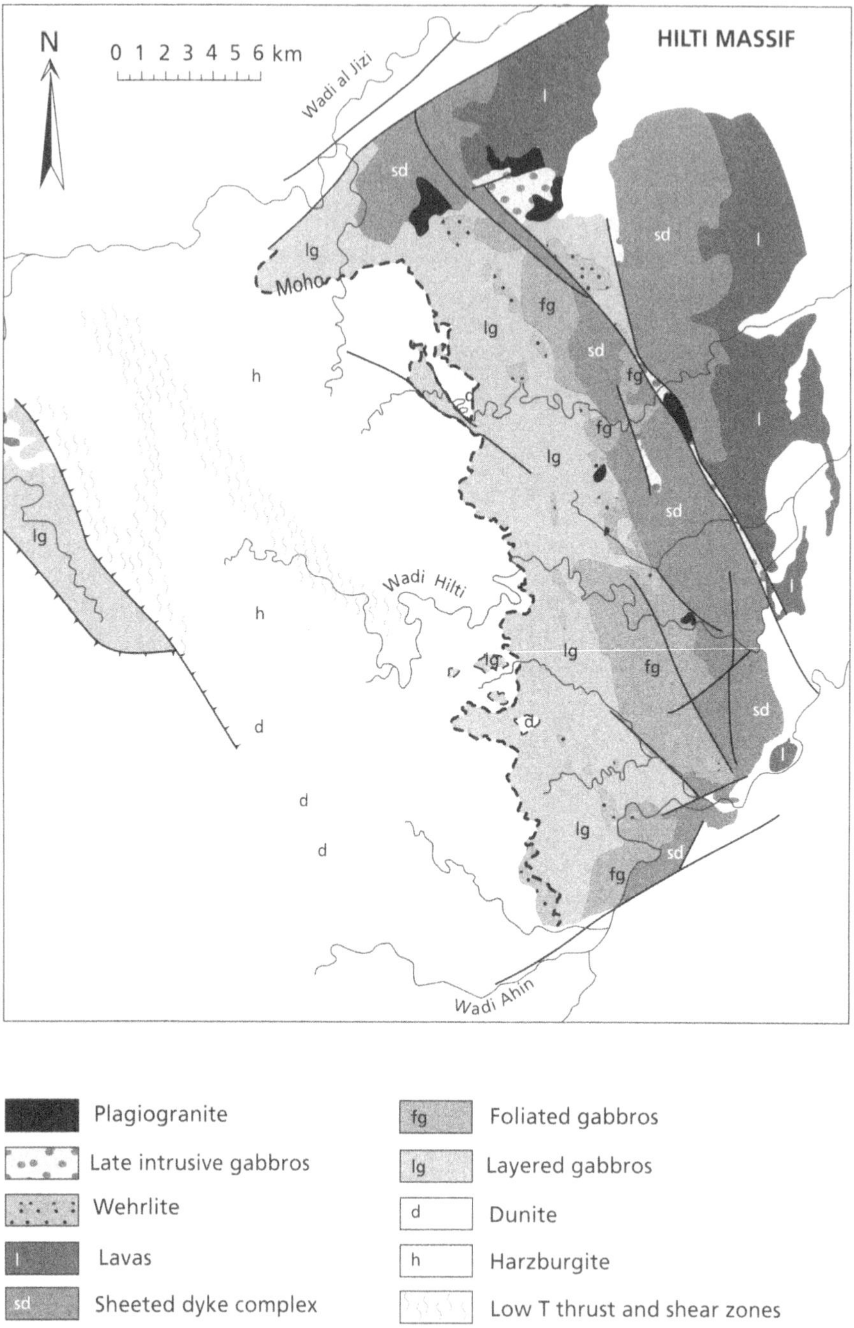

Figure 3. Geological map of Hilti massif, compiled from Glennie et al. (1974), MPM Oman (1987), Abrams et al. (1988), Ernewein et al. (1988) and Reuber (1991).

The Oman paleo-ridge

Numerous studies have been devoted to the Oman ophiolite because of its large dimension (500 x 80 km, Fig. 1), its remarkable preservation and its excellent exposure. All characteristics of the Oman ophiolite indicate that it originated at a fast-spreading oceanic ridge (Nicolas, 1989, p. 85; Nicolas et al., in press) and consequently it is expected that the structures which relate to ridge activity represent a steady-state functioning. Initiation of obduction by oceanic detachment occurred at the ridge itself or very close to the ridge. This has been shown (i) by structural analysis since detachment occurred in a hot and locally still active lithosphere (Boudier et al., 1988), and (ii) by radiometric dating since there is no significant age difference between the last stages of magmatic activity at the ridge and the first detachment-related basal metamorphism (Montigny et al., 1988; Hacker et al., 1993) . The data indicate that the detachment affected entire ridge segments, including diapirs probably scalped during their dying activity.

Hilti massif

The Hilti massif (Fig. 3) comprises, from E to W, all ophiolitic units: basalts, a well-developed sheeted dike complex which trends N-S and is subvertical to steeply west-dipping, a gabbro unit with a folded layering above a regular Moho surface, itself overlying a thin transition zone of about 10 to 100 m thickness and, finally, a vast domain of comparatively fresh peridotites of 4 to 5 km thickness at the most. The Hilti massif is a structurally homogeneous massif, on average inclined 25° eastward. This inclination is deduced from the average angle between sheeted dikes and their supposed original vertical orientation, and from the average dip of the uppermost mantle and lowermost gabbro foliations which are subparallel to the Moho.

The high-temperature foliations in the peridotites (Fig. 4) are fairly regular, on average parallel to the Moho surface, and the corresponding lineations are remarkably regular (Fig. 5 and 6), trending from E-W to WSW-ENE in the NW part of the massif. Like elsewhere in the Oman ophiolites, the trends of magmatic lineations in the layered gabbros are in continuity with the plastically developed lineations in the upper mantle rocks (Nicolas et al., 1988 ; Nicolas et al., in press), and the overall pattern of all of these lineations is arc-shaped (Fig. 6).

Along the western and lower contact, a comparatively low-temperature deformation, ascribed to the early detachment of the ophiolite (Boudier et al., 1988) , is superimposed on the high-temperature structures. This new foliation is also parallel to the other planar structures in the massif, suggesting that the detachment surface was subhorizontal when it developed

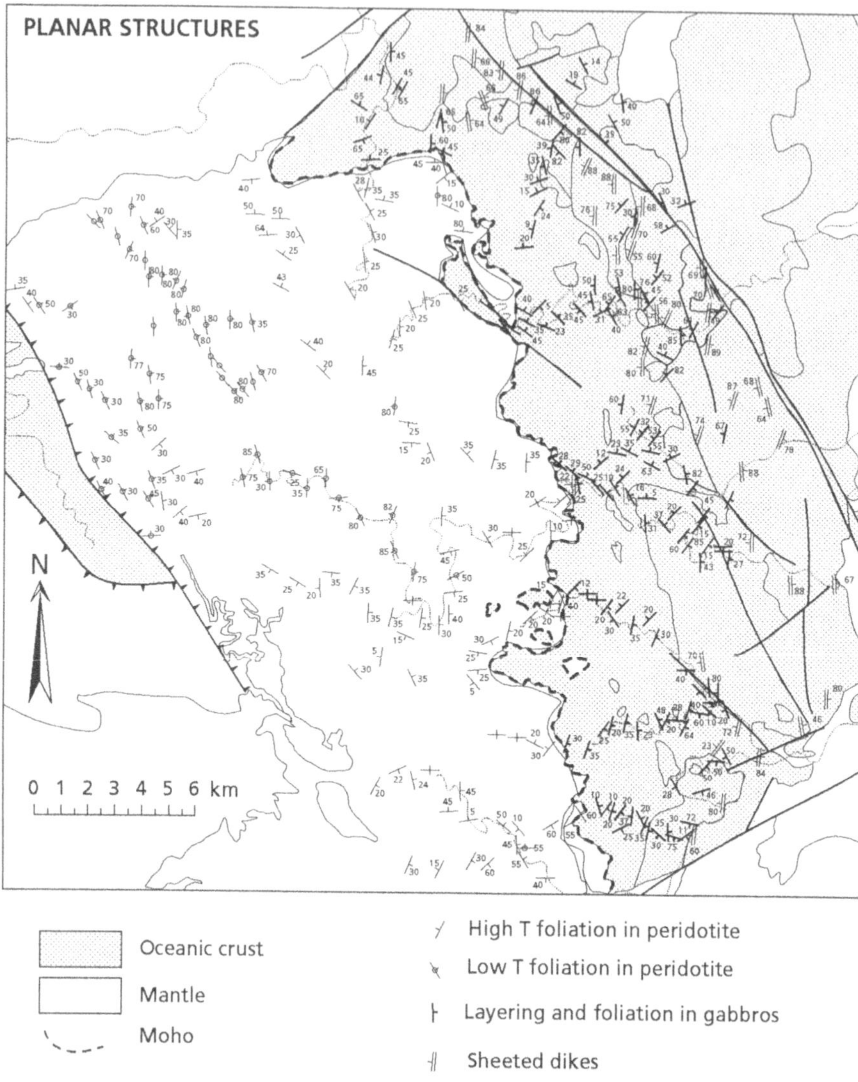

Figure 4. Map of the planar structures in the Hilti massif. Numbers indicate dips of foliations and dikes.

in the mantle of origin. The associated lineations trend more northerly than the high-temperature ones. Low-temperature to mylonitic deformation is also recorded along km-wide vertical shear zones (Figs. 3 to 7) which probably relate to the basal thrust, as shown for similar shear zones in the Fizh massif located just North of Hilti (Boudier et al., 1988) . These low-temperature areas are not further considered in the present study.

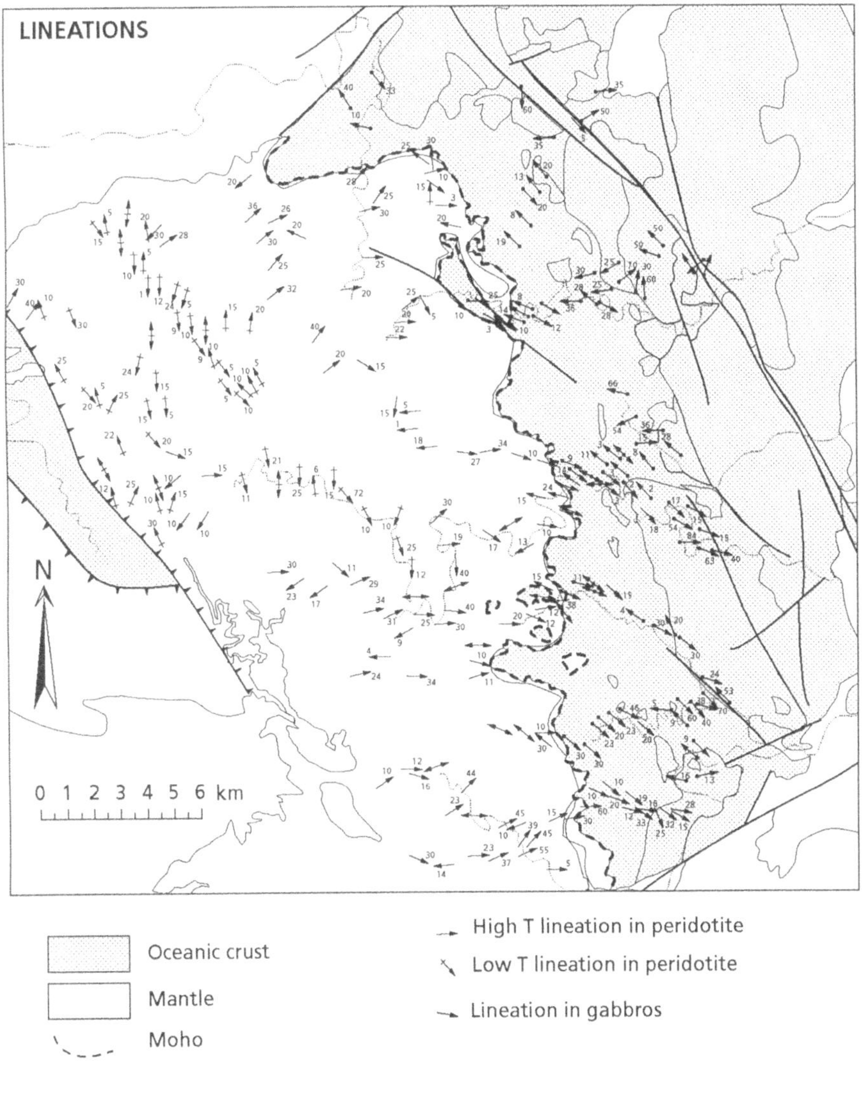

Figure 5. Map of the lineations in the Hilti massif, with numbers indicating plunges of the lineations. In the mantle peridotites, the mineral stretching lineations result from plastic flow and are best identified by elongated spinels and sometimes pyroxenes. In the gabbros, mineral lineations result from magmatic flow (traces of plastic strain are very scarce) and are defined by the preferred orientation of inequant plagioclase, pyroxene and olivine grains.

Shear direction analysis in the Hilti massif

Around 100 shear direction determinations have been made in the high-temperature peridotites of the Hilti massif, using mainly the asymmetry of

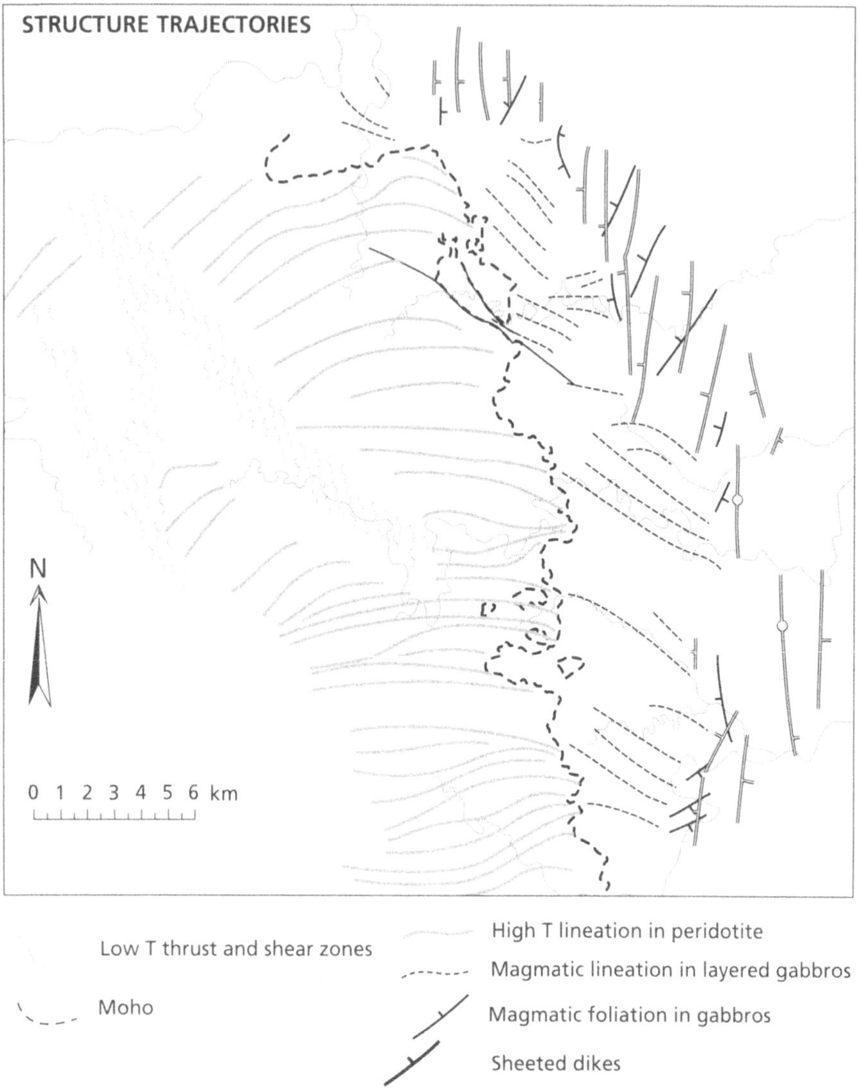

Figure 6. Map showing structure trajectories. See text for discussion.

olivine extinction with respect to the foliation trace in the XZ-plane of the deformation (i.e., the plane normal to the foliation and parallel to the lineation) and, where necessary, petrofabric measurements in olivine (see Nicolas and Poirier, 1976 , p. 333). The results are shown in Fig. 7. Around 20 shear determinations have been made in the magmatically deformed layered gabbros of the basal crustal section. These determinations are

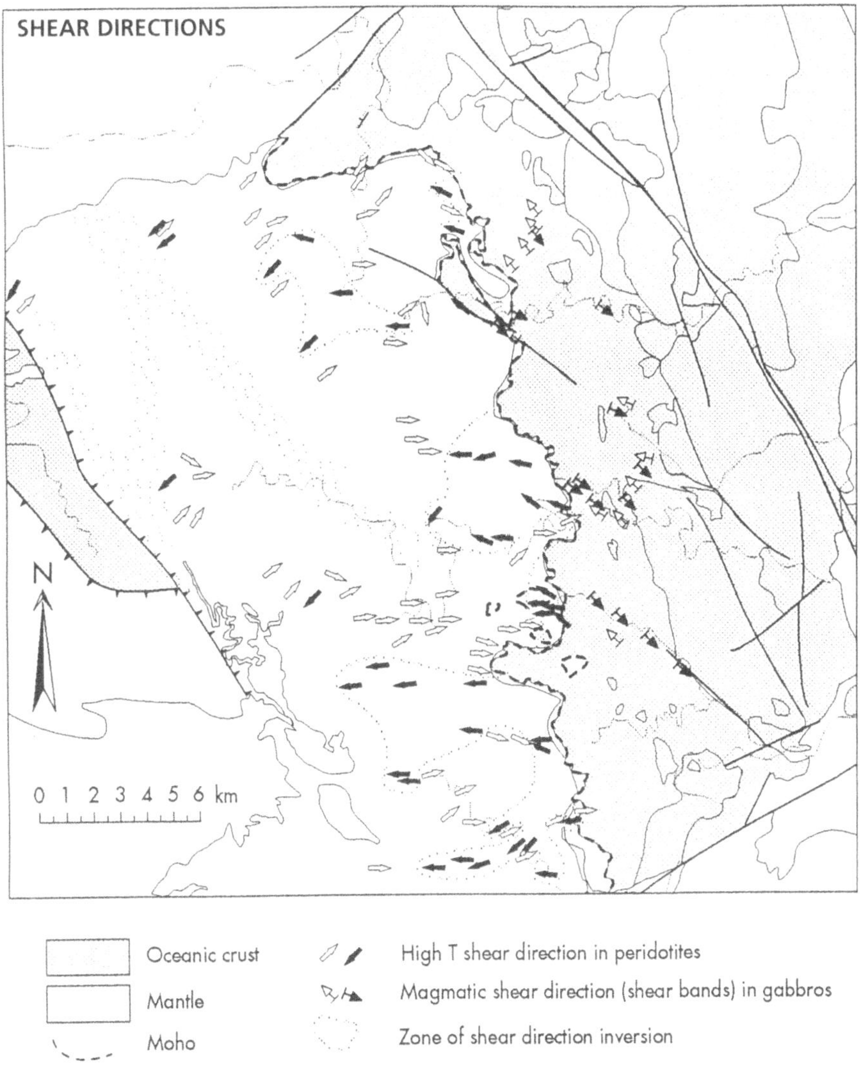

Figure 7. Map of the shear directions. Top-to-the-W and top-to-the-E movement directions are indicated by black and white arrows, respectively. For further discussion see text.

based on the shear direction of small magmatic shear bands (Nicolas, 1992) , which, in Oman, do not consistently dip in one and the same direction; counts of such shear bands often result in nearly equally partitioned, opposite shear directions, either at the outcrop or at the cross-section scale. It is, therefore, uncertain if the dominant shear directions in the gabbros shown on the map of Fig. 7 are all significant

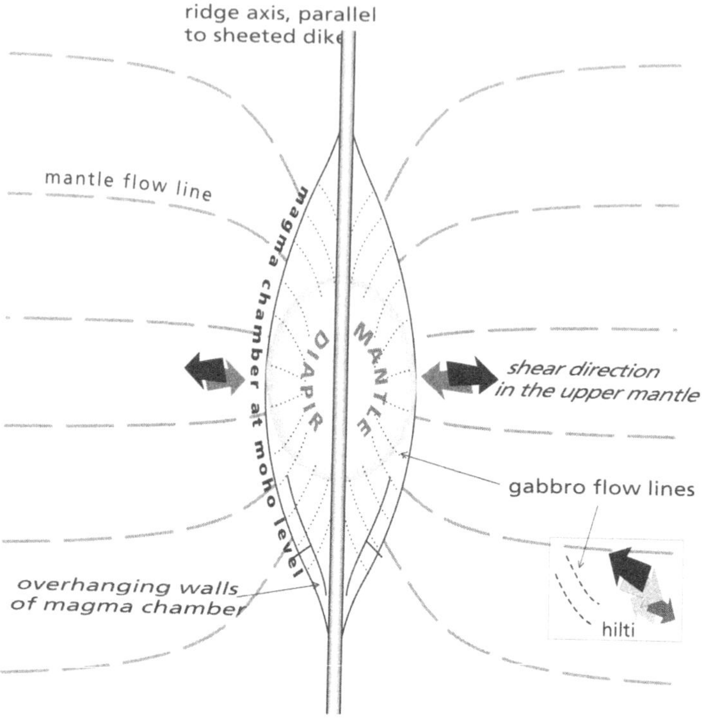

Figure 8. Schematic map of a ridge segment. The Hilti massif is interpreted to have been approximately located SE of the diapir. This interpretation is supported by the shear directions in the mantle, the obliquity between the upper gabbro foliations (parallel to the walls of the magma chamber) and the sheeted dikes, and the curvature of the lineation trajectories. See text for further discussion.

Discussion

Analysis of shear flow

Inspection of the shear directions in the mantle section (Fig. 7) shows, that there is no clear separation between domains with top-to-the-West and domains with top-to-the-East senses of movement (below referred to as westward and eastward directions, respectively). Eastward directions dominate in the western and deeper part of the massif whilst, in contrast, westward directions are at least as abundant as eastward ones in the eastern and shallower part of the ultramafic section. The few westward directions recorded in the western part of the massif may result from overprinting by low-temperature deformation with dominantly south-westward shear directions (Billiau, 1993) . A similar explanation cannot be invoked in the eastern, uppermost mantle section to explain a certain mixture of shear directions. However, starting from the Moho where the dominant shear direction is westward, it seems possible to define domains with this westward shear direction, extending westward between other domains dominated by eastward directions (Fig. 7). Such a distribution may result from the interplay between the topography and large-wavelength undulations of the Moho: because the Moho surface is locally horizontal (e.g., in the lower third part of the map) it is possible that domains with westward shear directions extending far into the peridotites may represent the uppermost mantle just below an eroded crustal unit. Alternatively, these domains may represent channels of westward shear flow within the underlying eastward domain.

In the magmatically deformed gabbros we observe, at the scale of the massif, a near-equal partitioning between westward and eastward directions (10 vs. 14), i.e., a similar partitioning as observed at the scale of individual exposures as outlined above. This confirms earlier observations from other massifs and is interpreted as indicating that the last magmatic flow recorded in the discrete shear bands (Nicolas, 1992) involved a dominantly coaxial flow regime (i.e., dominant flattening with a minor shear component only).

Location of the Hilti massif with respect to the paleo-ridge

The question of how to locate a given ophiolite with respect to its paleo-spreading centre has been discussed using structural and kinematic criteria (Nicolas, 1989, p. 10). Figures 8 and 9 summarise the various criteria that we favour in Oman and which have been tested in two massifs (Nicolas and Boudier, in press). We tentatively locate the Hilti massif at some distance and South-East of a diapiric centre (Fig. 8) using the following criteria:
- The presence of a thin transition zone: the transition zone just below the Moho is a zone of intense reaction between mantle peridotites and rising basaltic melts. It is mainly developed above mantle diapirs where it can attain thicknesses of around 700 metres, whilst it reduces to a few tens of metres away from diapirs (Boudier and Nicolas, in press). This thickness reduction results from an increasing deformation away from the diapir as well as from the expulsion of melt towards the overlying magma chamber or newly accreted crust. The thin transition zone observed in the Hilti massif is akin to a position away from the diapir.
- The attitude of the foliation in the upper gabbros relative to the sheeted dike complex: structural observations in Oman show that the foliation in the upper gabbros dips away from the centre of the magma chamber (Nicolas and Boudier, in press; Nicolas et al., in press). The more easterly trend of the foliations militate for the proposed model but the dip of these foliations is not clearly eastward as would be expected and a number of foliations have a westward dip with respect to local sheeted dikes (Fig. 4 and 6).

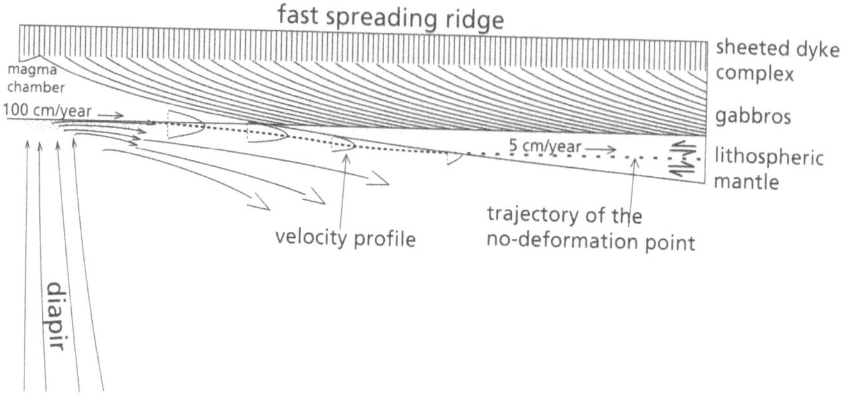

Figure 9. Schematic cross-section of a fast spreading ridge, perpendicular to the ridge axis. The forced flow expelled from the diapiric area is responsible for velocity profiles such that the maximum velocity corresponds to a point of no deformation, hence a shear direction inversion. Below the magma chamber this point is situated at the level of the Moho; as the mantle flows away from the diapir it moves further downward, and is finally accreted to the lithosphere when the flow is not forced anymore and where the lithosphere and asthenosphere velocities are equal. The diagram also illustrates the criteria (shear directions, obliquity of gabbro foliation versus sheeted dikes) used to locate the massif in the ridge reference frame.

- The trends of magmatic flow lines in the gabbros: the magmatic flow lines in the gabbro sequence show a more NW-SE oriented trend as compared with the more E-W trending flow lines in the mantle peridotites. In an off-axis situation, structures in the mantle are necessarily younger than the ones in the gabbros as the latter is accreted to the lithosphere while the mantle is still flowing (Fig. 9). The present curvature (Fig. 6) suggests the existence of a southward flow component in the magma chamber which would record the feeding of the southern extremity of this chamber. The NW-SE trend recorded in the gabbros is thus related to an early flow direction close to the diapiric centre (Fig. 8), whilst the E-W trend in the mantle relates to flow at some distance from the diapir, recorded when the mantle was accreted to the lithosphere (Fig. 8 and 9).

Mantle flow model

Assuming that the Hilti massif originated on the eastern flank of a NS-directed paleo-ridge (Fig. 8), the following conclusions can be derived.
- The dominant westward shear direction in the peridotites just below the Moho and eastward shear direction at greater depth are in agreement with the forced flow interpretation (outward shear direction below and inward shear direction above, Fig. 8 and 9).
- The limit between eastward and westward shear flow (point of no defor-

mation, Fig. 9) does not appear as a planar surface. It occurs at variable depth below the Moho, as shown by the E-W extension of some domains of westward flow (Fig. 7), suggesting that mantle flow is focussed along channels with comparatively more stagnant zones in between, although the extent of the westward shear domains, in particular the one enclosed in the northern part of the mantle unit, may also partly correspond to locally preserved remnants of the uppermost mantle. Depending on which interpretation is favoured to explain the depth variation of shear inversion, the westward shear domains may be interpreted either as relatively stagnant areas or as rapid flow channels. The flow localization may result either in an upward shift of the shear inversion towards the Moho (channel "grooving" into the overlying lithosphere; the westward shear domain would then correspond to the more stagnant regions) or in a downward shift of the shear inversion (the westward shear domains would then represent the focussed flow channels). The available data do not allow us to distinguish between the two hypotheses. No systematic differences have been recorded between the intensities of the linear fabrics in the westward and eastward shear domains, however, the distinct variations of the high-temperature foliation azimuth and dip beneath the westward shear domains (Fig. 7) may be in favour of the second interpretation (westward shear domains corresponding to rapid focussed flow channels).

- In contrast to areas located close to diapirs where the outward shear direction alone is recorded in the mantle section and where inward shear is supposed to take place within the large overlying magma chamber or very close to the Moho (Figs. 2, 8 and 9), the observation of a reverse flow in the mantle section of the Hilti massif tends to confirm its location further away from the feeding diapir. In this respect, the apparently larger extension of westward directions (inward relative shear flow) in the southern part of the Hilti massif as compared to the northern part is coherent with a location as proposed in Figure 8, with the northern part of the massif closer to the diapiric centre of origin than the southern part.

Conclusion

We have studied the shear directions associated with high-temperature plastic flow in upper mantle peridotites from the Hilti massif of the Oman ophiolite, in order to further investigate a shear direction inversion also found in other ophiolites. Such a shear direction inversion is not systematically present in Oman. The result of this investigation confirms the existence of a shear direction inversion in Hilti, with dominantly westward shear just below Moho, and eastward shear beneath. There is, however, a sharp contrast between the Hilti situation and that of diapiric areas in Oman such as Rustaq and Maqsad (Figs. 1 and 2). In the latter, the shear direction is exclusively top-away from the ridge (outward direction), whereas in Hilti the two directions exist and are somewhat mixed. This mixing of shear directions, unexpected in view of the remarkably regular structure of the Hilti massif, is tentatively explained by the localization of the flow expelled from the diapir, resulting in a variable depth of the shear inversion along strike. If this interpretation, which can be verified by further field research, is essentially correct, it would indicate that the forced flow, just below the Moho in the vicinity of an oceanic ridge and around the diapir, is not regular and parallel to the presumably planar lithosphere-asthenosphere boundary, but that it concentrates into channels, either locally moving the lithosphere-asthenosphere boundary upward (i.e. reducing the lithosphere thickness) or moving the shear inversion further down into the asthenosphere.

Acknowledgements

This paper benefited of careful reviews by G. Suhr and an anonymous referee. We acknowledge the help received from F. Boudier and E. Gnos who have participated in the field study of the Hilti massif. Previous data collected by G. Ceuleneer have been used and we thank him for providing his field book. This work was supported by the program DBT from the Centre National de la Recherche Scientifique. Mohammed Bin Hussain Bin Kassim, General Director of Minerals at the Ministry of Petroleum and Minerals of the Sultanate of Oman is kindly thanked for hospitality and help in Oman.

References

Abrams, M.J., Rothery, D.A. and Pontual, A., 1988. Mapping in the Oman ophiolite using enhanced thematic mapper images. Tectonophysics, 151: 387-401.

Billiau, S., 1993. Etude des écoulements asthénosphériques forcés au voisinage des diapirs mantelliques dans l'ophiolite d'Oman. Unpublished DEA, Université Montpellier II.

Boudier, F., Ceuleneer, G. and Nicolas, A, 1988. Shear zones, thrusts and related magmatism in the Oman ophiolite : initiation of thrusting on an oceanic ridge. Tectonophysics, 151: 275-296.

Boudier, F. and Nicolas, A., in press. Nature of the Moho transition zone in the Oman ophiolite. J. Petrol.

Cassard, D., 1980. Structure et origine des gisements de chromite du massif du Sud (ophiolite de Nouvelle Calédonie). Guides de prospection. Unpublished Thèse de 3° cycle, Université de Nantes.

Ceuleneer, G., 1991. Evidence for a paleo-spreading center in the Oman ophiolite: mantle structures in the Maqsad area. In: T. Peters, A. Nicolas and R. G. Coleman (Editors), Ophiolite genesis and evolution of the oceanic lithosphere. Kluwer Academic Publishers, Dordrecht, pp. 147-173.

Ceuleneer, G., Nicolas, A. and Boudier, F., 1988. Mantle flow patterns at an oceanic spreading centre: the Oman peridotites record. Tectonophysics, 151: 1-26.

Ernewein, M., Pflumio, C. and Whitechurch, H., 1988. The death of an accretion zone as evidenced by the magmatic history of the Sumail ophiolite. Tectonophysics, 151: 247-274.

MPM Oman, 1987. Geological map, Al Wasit, NG40-14E-III. Ministry of Petroleum and Minerals, Sultanate of Oman.

Girardeau, J. and Nicolas, A., 1981. Structures in two of the Bay of Islands (Newfoundland) ophiolite massifs: a model for oceanic crust and upper mantle. Tectonophysics, 77: 1-34.

Glennie, K.W., Boeuf, M.G.A., Hughes Clarke, M.W., Moody-Stuart, M., Pilaar, W.F.H. and Reinhardt, B.M., 1974. Geology of the Oman mountains". Verh. Koninklijk Nederlands Geologisch Mijnbouwkundig Genootschap, 31.

Hacker, B.R., McWilliams, M.O. and Mosenfelder, J.L.., 1993. Rapid emplacement of the Oman ophiolite. IAVCEI Abstracts, Ancient volcanism and modern analogues, Canberra, p. 44.

Hoxha, M., 1993. Etude structurale et pétrologique de l'ophiolite de Kukes (Albanie): cinématique de la déformation et géométrie de la ride. Unpublished Thesis INPL, Université de Nancy.

Montigny, R., Le Mer, O., Thuizat, R. and Whitechurch, H., 1988. K-Ar and 40Ar/39Ar study of metamorphic rocks associated with the Oman ophiolite: tectonic implications. Tectonophysics, 151: 345-362.

Nicolas, A., 1989. Structures in ophiolites and dynamics of oceanic lithosphere. Kluwer Academic Publishers, Dordrecht, 367 pp.

Nicolas, A., 1992. Kinematics in magmatic rocks with special reference to gabbros. J. Petrology, 33: 891-915.

Nicolas, A. and Boudier, F., in press. Mapping mantle diapirs and oceanic crust segments in Oman ophiolites. J. Geophys. Res.

Nicolas, A., Boudier, F. and Ildefonse, B., in press. Evidence from the Oman ophiolite for active mantle upwelling beneath a fast-spreading ridge. Nature.

Nicolas, A. and Poirier, J.P., 1976. Crystalline plasticity and solid-state flow in metamorphic rocks. Wiley-Interscience, London, 444 pp.

Nicolas, A., Reuber, I. and Benn, K., 1988. A new magma chamber model based on structural studies in the Oman ophiolite. Tectonophysics, 151: 87-105.

Nicolas, A. and Violette, J.F., 1982. Mantle flow at oceanic spreading centers: models derived from ophiolites. Tectonophysics, 81: 319-339.

Podvin, P., 1983. Remobilisations chimiques successives dans les tectonites ophiolitiques et leurs gisements de chromite. Exemple du massif du Humboldt, Nouvelle Calédonie. Unpublished thèse Doc. Ing., Ecole Nat. Sup. Mines, Paris.

Prinzhofer, A., Nicolas, A., Cassard, D., Moutte, J., Leblanc, M., Paris, P. and Rabinovitch, M., 1980. Structures in the New Caledonia peridotite-gabbros: implications for oceanic mantle and crust. Tectonophysics, 69: 85-112.

Rabinowicz, M., Ceuleneer, G. and Nicolas, A., 1987. Melt segregation and flow in mantle diapirs below spreading centers : evidence from the Oman ophiolite. J. Geoph. Res., 92: 3475-3486.

Rabinowicz, M., Nicolas, A. and Vigneresse, J.L.., 1984. A rolling mill effect in asthenosphere beneath oceanic spreading centers. Earth Planet. Sci. Letters, 67: 97-108.

Reuber, I., 1991. Geometry and flow pattern of the plutonic sequence of the Salahi massif (Northern Oman Ophiolite) - A key to decipher successive magmatic events. In: T. Peters, A. Nicolas and R. G. Coleman (Editors), Ophiolite genesis and evolution of the oceanic lithosphere. Kluwer Academic Publishers, Dordrecht, pp. 83-103.

Suhr, G., 1992. Upper mantle peridotites in the Bay of Islands ophiolite, Newfoundland: formation during the final stages of a spreading centre? Tectonophysics, 206: 31-53.

Violette, J.F., 1980. Structure des ophiolites des philippines (Zambales et Palawan) et de Chypre. Ecoulement asténosphérique sous les zones d'expansion océaniques. Unpublished Thèse de 3° cycle, Université de Nantes.

Part III

Numerical Modelling

Non Steady-State Thermal Model of Spreading Ridges: Implications for Melt Generation and Mantle Outcrops

CHANTAL TISSEAU AND THIERRY TONNERRE

CNRS URA 1278 Domaines Océaniques, Département des Sciences de la Terre, Université de Bretagne Occidentale, BP 809, 29285 Brest cedex, France

Abstract

A non steady-state thermal model is proposed for the axial domain of spreading ridges, in which accretion is simulated as the superposition of seafloor spreading and thermal inputs which vary through time associated with magmato-tectonic cycles. Realistic time frequencies of accretion are tested for extreme cases of slow and fast spreading ridges. The thermal structure, melt fraction and crustal production are computed. For fast spreading ridges, the partial melt region is very wide and varies little through time; the associated crustal production is large and nearly constant. In contrast, melting zones beneath slow spreading ridges are narrow and waxing and waning through time; the associated crustal production decreases dramatically toward zero at the end of cycles. Our model reconciles opposing views on the geometry of the melting regions beneath spreading centres and their relationship to spreading rate. In addition, it accounts for the occurrence of mantle rock outcrops reported along the ridge axis and for the relationship between their relative abundance and axial segmentation.

Introduction

Outcrops of mantle rocks on the seafloor have been frequently observed along the mid-ocean ridge system. However, the emplacement of these mantle rocks is not randomly distributed along the ridge axis, revealing a discontinuous magmatic crust which relates closely to axial segmentation defined by geological and geophysical data. These variations in magmatic crustal thickness are currently interpreted in terms of magma supply variations. Magma and heat supplies are closely linked, as magma supply both affects and reflects heat supply. It follows that crustal thickness variations

R.L.M. Vissers and A. Nicolas (Eds.),
Mantle and Lower Crust Exposed in Oceanic Ridges and in Ophiolites, 181-214.

Table 1. Observations of mantle rock exposures at the ends of ridge segments. FZ: Fracture Zone, MAR: Mid-Atlantic Ridge.

Locations		*Full rate (mm/yr)*	*Description*	*References*
Unnamed FZ	MAR/79°N			Michael and Bonatti, 1985a
Gibbs FZ	MAR/53°N	26	North and south walls	Hékinian and Aumento, 1973 Michael and Bonatti, 1985a Dick, 1989
Unnamed FZ	MAR/43°N			Phillips et al., 1969 Thompson and Melson, 1972
B FZ	MAR/36°30'N	22		Dick, 1989
Oceanographer FZ	MAR/35°N	26	East intersection	OTTER, 1984 Michael and Bonatti, 1985a
Atlantis FZ	MAR/30°N	26	North wall and near top of crestal mountain	Miyashiro et al., 1969 Michael and Bonatti, 1985a
Kane FZ	MAR/24°N	30	North and south walls	Miyashiro et al., 1969 Michael and Bonatti, 1985a
			North and south walls	Dick, 1989
Fifteen Twenty FZ	MAR/15°20'N	22	East intersection	Rona et al., 1987
Vema FZ	MAR/10°N	24		Bonatti et al., 1971 Thompson and Melson, 1972
			North and south walls	Honnorez and Kirst, 1975
			South wall	Prinz et al., 1976
			North and south walls	Bonatti and Honnorez, 1976 Michael and Bonatti, 1985a
			Deepest part of transform valley and south wall	Lagabrielle et al., 1992
St Paul FZ	MAR/0°S		West and east intersections	Bonatti et al., 1971 Thompson and Melson, 1972
Romanche FZ	MAR/2°S	33	North wall	Melson and thompson, 1970
			North and south walls	Bonatti et al., 1971 Thompson and Melson, 1972
			North and south walls	Honnorez and Kirst, 1975
			North and south walls	Prinz et al., 1976
			North and south walls	Bonatti and Honnorez, 1976 Michael and Bonatti, 1985a
Chain FZ	MAR/3°S		North wall	Bonatti et al., 1971
Unnamed FZ	MAR/4°S		South wall	Thompson and Melson, 1972

may indicate a discontinuous contribution of heat beneath the ridge axis, which leads to variations in the thermal structure. Furthermore, a recent review by Cannat (1993) emphasizes the role of a thick axial lithosphere in the emplacement of mantle rocks on the seafloor at oceanic ridges. The axial lithospheric thickness is directly deduced from the thermal structure beneath the ridge axis.

Observations regarding the emplacement and abundance of mantle rocks at oceanic ridges can, therefore, be related to the thermal regime via the variations in partial melting and crustal production. In contrast to

Table 1 continued. Observations of mantle rock exposures at the ends of ridge segments. FZ: Fracture Zone, CIR: Central Indian Ridge, SWIR: South West Indian Ridge, AAR: American Antarctic Ridge, EPR: East Pacific Rise

Locations		*Full rate (mm/yr)*	*Description*	*References*
Owen FZ	CIR/10°N		40 km SW of intersection with Carsberg Ridge West wall	Hamlyn and Bonatti, 1980 Bonatti et al., 1983
Vema FZ	CIR/9°S	32	North and south walls	Engel and Fisher, 1975
Argo FZ	CIR/13°30'S	36	North and south walls Deepest part of transform valley	Engel and Fisher, 1975
Marie Celeste FZ	CIR/17°30'S	42	North wall, deepest part of transform valley	Engel and Fisher, 1975
Bouvet FZ	SWIR/1°E	14	West wall	Dick, 1989
Islas Orcadas FZ	SWIR/6°E	14	South-east wall	Dick, 1989
Shaka FZ	SWIR/9°E		East wall	Dick, 1989
S FZ	SWIR/14°E	14.4	South wall	Dick, 1989
DuToit FZ	SWIR/25°30'E	16		Dick, 1989
Bain FZ	SWIR/32°E			Dick, 1989
P. Edward FZ	SWIR/35°E	16		Dick, 1989
Discovery FZ	SWIR/42°E			Dick, 1989
46°E FZ	SWIR/46°E	14.8		Dick, 1989
Atlantis II FZ	SWIR/57°E	14.5	Median tectonic ridge East and west walls	Johnson and Dick, 1992
Melville FZ	SWIR/62°E	16	West wall South at 10-24.5 Ma	Engel and Fisher, 1975 Bassias and Triboulet, 1992
Bullard FZ	AAR/7°W	18	South-east wall	Dick, 1989
Vulcan FZ	AAR/16°W	18	North wall	Dick, 1989
Garret FZ	EPR/13°28'S	180	Deepest part of transform valley	Hébert et al., 1983 Cannat et al., 1990a Bideau et al., 1991 Hékinian et al., 1992
Iti FZ	EPR/24°13'S	120-150		Constantin et al., 1993
Heezen FZ	EPR/55°S	88		Lonsdale, 1986

the large-scale thermal evolution of the oceanic lithosphere, the thermal regime in the vicinity of the ridge axis is relatively poorly known. In particular, no current model incorporates a possible variability with time of accretionary processes at the ridge axis. Therefore, we have developed a thermal model in the non-steady state regime more appropriate to a study of the axial domain. Our aim in this paper is to quantify the thermal structures which may be expected beneath the axes of slow and fast spreading ridges, and explore their implications for partial melting, i.e., for the presence or absence of a melt zone, the dimensions of this melt

Table 2 Observations of mantle rock exposures inside ridge segments. MAR: Mid-Atlantic Ridge, FZ: Fracture Zone, CIR: Central Indian Ridge, SWIR: South West Indian Ridge, EPR: East Pacifid Rise.

Locations	*Full rate (mm/yr)*	*Description*	*References*
MAR/39°N	20	DSDP site 556, west at 35 Ma	Michael and Bonatti, 1985a and b Juteau et al., 1990b
MAR/37°N	22	DSDP site 334, west at 8.9 Ma	Shipboard Scientific Party, 1977 Hodge and Papike, 1977 Juteau et al., 1990b Girardeau and Francheteau, 1993
		DSDP site 558, west at 37 Ma	Michael and Bonatti, 1985a and b Juteau et al., 1990b
MAR/34°N	26	DSDP site 560, west at 12 Ma	Michael and Bonatti, 1985a and b Juteau et al., 1990b
MAR/26°N	26	East at 3 Ma	Tiezzi and Scott, 1980 Michael and Bonatti, 1985a
MAR/22°45'N-24°N	30	West wall, 30 km south of Kane FZ	Karson et al., 1986 and 1987 Mével et al., 1988 and 1991 Karson, 1991
		ODP site 670, west wall, 45 km south of Kane FZ	Karson et al., 1986 and 1987 Shipboard Scientific Party, 1988 Cannat et al., 1990b Juteau et al., 1990a and b Karson, 1991
		DSDP site 395, west, 6.5 to 7.2 Ma, 90 km south of Kane FZ	Arai and Fujii, 1979 Boudier, 1979 Sinton, 1979 Michael and Bonatti, 1985a Juteau et al., 1990b
MAR/16°52'N	26	East wall	Cannat et al., 1992
MAR/15°37'N	26	Top of the east wall and west wall	Cannat et al., 1992
MAR/6°N		Pillsbury P6903-28, east (?) wall	Bonatti et al., 1971 Bonatti et al., 1975 Michael and Bonatti, 1985a
CIR/12°25'S	36	East wall	Engel and Fisher, 1975
SWIR/7°34'E	14	Wall	Dick, 1989
SWIR/13°076'E	14.4	Small rift valley high	Dick, 1989
Cayman R/18°N and 18°20'N	20	East and west walls	CAYTROUGH, 1979 Stroup and Fox, 1981 Malcom, 1981 Ito and Anderson, 1983
Mathematician R/17°N	< 43		Vanko and Batiza, 1982
EPR/02°15'N	130-135	Hess deep, North wall, at 1Ma	Rudnik, 1976 Kashintsev et al., 1982 Francheteau et al., 1990 Girardeau and Francheteau, 1993 Hékinian et al., 1993

zone, and for the range of possible values for the melt fraction and associated crustal production.

Mantle rock outcrops at ridges: relationship to axial segmentation.

Most outcrops of serpentinized peridotites at spreading ridges are described from the ends of ridge segments in fracture zone settings. This observation was first made in the seventies by several authors including Thompson and Melson (1972), Engel and Fisher (1975), and Bonatti and Honnorez (1976). More recently, Dick (1989) has discussed the presence of peridotites along very slow spreading ridges. Using available geological and geophysical data on axial outcrops of mantle rocks, Cannat (1993) has constrained their emplacement conditions. In this paper we build on these previous studies, emphasizing the relationship between peridotite distribution and axial segmentation.

Observations

The distribution and abundance of mantle rock outcrops strongly depend not only on spreading rate but also on position within ridge segments. Most of the documented outcrops of peridotites are located along slow spreading ridges, i.e., the Mid-Atlantic Ridge (M.A.R.), the South-West Indian Ridge (S.W.I.R.), and the Atlantic-Antarctic Ridge (A.A.R.), and only very few have been observed along fast spreading ridges such as the East-Pacific Rise (E.P.R.). Tables 1 and 2 summarize the known peridotite outcrops at ridges. Peridotites exposed at the ends of ridge segments, referred to in this paper as "segment boundary" peridotites, are listed separately (Table 1) from those located within a ridge segment, i.e., clearly away from discontinuities along the ridge axis. Such peridotites, referred to below as "intra-segment" peridotites, are listed in Table 2.

Along slow spreading ridges, intra-segment peridotites do occur but they are much less common than at the ends of segments. In both cases, i.e., in intra-segment and in segment-boundary settings, outcrops are located on the valley walls as well as on the inner floors of the axial or transform valleys (Tables 1 and 2). The relative abundance of peridotites in intra-segment and segment-boundary settings are also reflected in the relative proportion of peridotites in dredge hauls collected at slow spreading ridges: peridotites are markedly less abundant at intra-segment localities than along the fracture zones where peridotites commonly make up the majority of the samples (Dick, 1989).

Except for the case north of Hess Deep, no peridotite outcrops have been observed at intra-segment sites of fast spreading ridges, and only a few have been found in the associated fracture zones (Table 1). These latter occurrences are located in the deepest part of the deepest E.P.R. fracture zones, commonly interpreted as potential cross-sections of the oceanic upper lithosphere. It is noted that the occurrence of peridotites in Hess Deep seems to be principally associated with the westward Cocos-Nazca ridge propagation, rather than with normal accretion processes (Francheteau et al., 1990).

We may conclude that the abundance of peridotites exposed on the seafloor at ocean ridges decreases with increasing spreading rate, and that this abundance may become zero on fast spreading ridges. In addition, at slow as well as fast spreading ridges, segment-boundary peridotites are much more abundant than intra-segment peridotites.

Interpretations

What kind of processes may emplace peridotites at oceanic ridges? Several authors (Dick et al., 1981; Francheteau et al., 1990; Mével et al., 1991; Cannat, 1993) propose that these outcrops result from tectonic denudation of the upper crust through low-angle detachment faulting. The likely occurrence of a shear zone beneath the M.A.R. near 23°N has been suggested by Toomey et al. (1988) from a micro-earthquake experiment in the axial valley. Such a model would predict a pronounced asymmetry in the surface data. There is, indeed, an asymmetry in the topography near ridge discontinuities, but the rift valley is usually symmetric near segment mid-points (Forsyth, 1991; Sempéré et al., 1993). Tectonic denudation may explain the occurrence of mantle rocks in the deepest parts of the fracture zones as reported on the fast spreading E.P.R. but the peridotites located on top or along entire walls of both the transform and axial valleys of slow spreading ridges remain unexplained by this model.

An alternative interpretation is that the deep rock outcrops may occur in regions of lowest melt production, revealing variations in magma supply along the length of a segment (Dick, 1989; Cannat, 1993; Sempéré et al., 1993). Juteau et al. (1990b) noted that the emplacement of deep rocks on the seafloor, away from fracture zones, is a characteristic feature of slow spreading centres. These authors link limited magma supply and variability of the magmato-tectonic cycle with the occurrence of purely tectonic phases of spreading: these phases are characterized by a total absence of magmatic construction and seem to occur repeatedly or last during long periods of time. Magma supply variations, both in space and time, have clear implications for the thermal regime beneath the ridge, hence for the extent of the melting zone and crustal production. At this stage, we wish to underline that both types of interpretation, tectonic denudation and limited magma supply, are related to a colder thermal state of the lithosphere.

The variability of accretionary processes: constraints on thermal modelling

In order to compute the thermal regime beneath oceanic ridges, we must constrain the simulation of the accretion. The main questions appear to be the following: are accretionary processes the same or not along an entire segment? And, at time-scales of the order of some thousands to some hundreds of thousands of years, are they synchronous or not? Variations in the accretion processes themselves would imply variations in the thermal regime, and therefore lead to different melt fractions and a variable crustal production along a segment as described above.

Seafloor spreading: a nearly two-dimensional and time-regular phenomenon

The fundamental evidence for accretion is the spreading of oceanic crust away from the ridge axis as revealed by the well-known linear pattern of magnetic anomalies. Off-axis magnetic isochrons record the temporal evolution of spreading at the ridge axis and, in spite of slight undulations observed in highly detailed surveys (Carbotte and Macdonald, 1992; Sloan and Patriat, 1992), they remain near-parallel to the axis. Seafloor spreading can thus be considered as a time-regular phenomenon and even, at large scales, as a steady-state phenomenon. This assumption underlies most existing models of accretion.

Spreading on oceanic ridges can be assumed to be similar to their subaerial counterparts where smaller-scale details of spreading can be studied. In studies of the Asal rift of East Africa, Ruegg et al. (1984) and Stein et al (1991) describe a seismic and magmatic activity focussed along an elongated narrow band of 5 to 10 km wide. Chadwick et al. (1991) report an event at the Juan de Fuca Ridge, during which an important volume of basalt has been emplaced at the axis along a long and narrow (1 km wide) band. Similarly, on the basis of side scan sonar data on the E.P.R. near 13°N, Vaslet (1993) has shown that the active deformation occurs within a very narrow, 1 km wide band. These studies indicate that emplacement of oceanic crust occurs within an elongated and narrow axial domain, without significant spatial variations along a segment, at least for intermediate to fast spreading rates.

Along-strike variability and ridge segmentation: variable heat supply beneath the ridge axis

In contrast with the foregoing, there are many cases of marked spatial variations within ridge segments or from one segment to another. Axial morphology in particular shows dramatic variations between shallow large domes and deep wide grabens. Morphology seems to principally depend on spreading rate, and the simple fast/slow spreading ridge classification

(Menard, 1967) is still in common use. But even away from any hotspot influence, a growing number of exceptions have been found to the direct and simple relationship between surface ridge expression and spreading rate. Table 3 lists some of these exceptions, and shows that both morphological types may coexist without any significant variation in spreading rate.

Most recent studies on oceanic ridges invoke a relationship with mantle temperature beneath the ridge to explain along-axis variations. Petrological studies (Whitehead et al., 1984; Klein and Langmuir, 1987; Dick, 1989), as well as micro-earthquake experiments (Toomey et al., 1985, 1988; Kong et al., 1992; Barclay et al., 1993) suggest hotter mantle material underneath the centres of individual ridge segments than underneath the ends. This implies that the segmentation is related to the underlying thermal structure. Gravity studies over slow spreading ridges (Kuo and Forsyth, 1988; Lin et al., 1990; Morris and Detrick, 1991; Neumann and Forsyth, 1993; Rommevaux et al., 1994) reveal a bull's eye pattern of gravity anomalies that overlie morphological segments. These data indicate a larger crustal thickness at the centres of ridge segments than at segment ends, and support the idea of a thermal anomaly linked to each segment. Work on ophiolites (Ceuleneer et al., 1988; Nicolas, 1989) is also consistent with the concept of hot mantle diapirs topped by a thick crust. On the other hand, bull's eye gravity patterns are not clearly observed on fast ridges (Madsen et al., 1990). Instead, the data suggest underlying elongated structures with only minor variations in axial crustal thickness (Lin and Phipps Morgan, 1992). Other authors propose a pattern of diapiric mantle upwelling beneath all ridges. For fast ridges, the axial domain would be hot enough to allow the along-axis flow of material away from the upwelling centres, either in the crust (Bell and Buck, 1992) or in the mantle (Dick, 1989), or in both (Nicolas, 1989). The relationship between surface segmentation and underlying mantle structure is still subject of debate, and we believe that more experimental constraints are needed to better understand the deep structure of ridges. It seems clear, however, that the thermal regime of the mantle plays a key role in this relationship.

A major problem related to the accretion process concerns any variability with time. Along-axis variations are also often explained in terms of periodicities in magma supply, implying that magmatic and tectonic processes at the ridge axis alternate through time. Assuming that magma supply coincides with heat supply beneath the ridge as suggested by Dick (1989) and Cannat (1993), such magmato-tectonic cycles are thus related to heat supply variations. The interpretation in terms of spatial variations between segments can be considered in terms of variability with time: the simplest explanation is that the thermal anomaly beneath each segment evolves with time instead of being a steady-state phenomenon. This view is in agreement with the across-axis variability observed in the morphology (see references in Table 4) or in crustal thickness as determined from residual gravity anomalies (Kuo and Forsyth, 1988; Morris and Detrick, 1991;

Table 3. Exceptions to the simple spreading rate/morphology rule (away from hotspot influence). MAR: Mid-Atlantic Ridge, SWIR: South West Indian Ridge, AAD: Australian Antarctic Discordance, PAR: Pacific Antarctic Ridge, JdFR: Juan de Fuca Ridge, EPR: East Pacific Rise

Locations	*Full rate (mm/yr)*	*Description*	*Methods*	*References*
MAR/26°S	34-38	dome	bathymetry	Batiza et al., 1988 Blackman and Forsyth, 1991 Grindlay et al., 1992
MAR/33°S	36	dome	bathymetry	Kuo and Forsyth, 1988 Neumann and Forsyth, 1993 Tolstoy et al., 1993
SWIR/120°-128°E (AAD)	74	grabens	bathymetry	Weissel and Hayes, 1974 Forsyth et al., 1987 Palmer et al., 1993
PAR/65°S	70	domes and grabens	altimetry	McAdoo and Marks, 1992 Sandwell, 1992 Small and Sandwell, 1992 Géli et al., 1994a
			bathymetry	Marks and Stock, 1994
JdFR/48°N	58	domes and grabens	bathymetry	Karsten et al., 1986 Davis et al., 1987
EPR/23°N	58	domes and grabens	bathymetry	Lewis, 1979 McClain and Lewis, 1980
EPR/16°N	96.5	domes and grabens	bathymetry	Langmuir et al., 1990
Chile Ridge/45°S	60	domes and grabens	bathymetry	Klein et al., 1993
Central North-Fiji Basin	60-80	domes and grabens	bathymetry	Auzende et al., 1988
Spreading Ridge				Lafoy et al., 1990 Gracia et al., 1994 Auzende et al., In press

Rommevaux, 1994). The durations of these magmato-tectonic cycles (according to different authors and presented in Table 4) range from 50,000 yr for the fastest ridges to 1 Ma for the slowest ones.

Spreading and heat supply

Although the thermal anomaly pattern and seafloor spreading are certainly connected, the relationship between these is not unequivocal. Recent highly detailed geophysical mapping of slow spreading ridges, such as the M.A.R. at 27-30°N (Lin et al., 1990; Sloan and Patriat, 1992; Rommevaux et al., 1994), at 20-24°N (Schulz et al., 1988; Morris and Detrick, 1991; Gente et al., in press) or at 26°S (Blackman and Forsyth, 1991; Grindlay et al., 1992), the Mohns Ridge (Géli, 1993; Géli et al., 1994b), and the Carlsberg Ridge (Rommevaux, 1994), reveal superpositions of mantle Bouguer anomaly and linear magnetic anomaly patterns as if there were no connection between them.

Table 4. Duration of magmato-tectonic cycles (Ma). EPR: East Pacific Rise, MAR: Mid-Atlantic Ridge, FAMOUS: French American Mid Ocean Undersea Study, AMAR: Alvin Mid-Atlantic Ridge, JdFR: Juan de Fuca Ridge, SWIR: South West Indian Ridge, SEIR: South East Indian Ridge

References	*Slow*	*Intermediate spreading ridge*	*Fast*	*Methods*	*Locations*
Lewis, 1979		0.3		Morphology (across and along axis)	EPR/23°N 68 mm/yr
Crane and Ballard, 1981	0.64			Morphology (along axis)	FAMOUS MAR/ 36°40'N-36°55'N and AMAR MAR/ 36°25'N, 20.4 mm/yr
Lichtman and Eissen, 1983		0.01 to 0.1		Kind of volcanism (geochronology)	JdFR and EPR/21°N 60 mm/yr
Courtillot et al., 1984	1			Field study (geochronology)	Afar, 16 mm/yr
Patriat and Courtillot, 1984	1			Triple junction stability (through time)	Bouvet triple junction and the SWIR 14 mm/y
Kappel and Ryan, 1986		0.25		Morphology (across axis)	JdFR, 60 mm/yr
Gente, 1987	0.5 to 1		0.05 to 0.1	Morphology (across and along axis)	review (MAR and EPR)
Pockalny et al., 1988	0.3 to 1			Morphology (along fracture zone)	Kane, MAR/24°N 30 mm/yr
Malinverno and Pockalny, 1990	0.7		0.07	Morphology (across and along axis)	review (MAR and EPR)
Sauter et al., 1991		0.26		Morphology (across axis)	SEIR/27°40'S, 62 mm/yr
Lagabrielle et al., 1992	0.3			Field study (within Ridge-Transform Intersection Domain)	Vema, MAR/10°N 24 mm/yr

Apart from the spreading, another phenomenon affects the axial domain. Although the overall regime may be the same for the entire ridge as suggested by the nearly two-dimensional, time-regular spreading, individual segments are not necessarily all at the same stage of their thermal evolution. Neighbouring segments seem to work independently, as already suggested

Table 5. Computational runs

	Slow spreading ridge		*Fast spreading ridge*	
Run number	1	2	3	4
Spreading	1 km each 0.05 Ma	1 km each 0.05 Ma	1 km each 0.005 Ma	1 km each 0.005 Ma
Magmato-tectonic cycle duration	1 Ma	0.7 Ma	0.07 Ma	0.05 Ma
Cycle/spreading ratios	1/0.05=20	0.7/0.05=14	0.07/0.005=14	0.05/0.005=10

by, e.g., Harper (1985), Crane and Ballard (1981), Grindlay et al. (1992), Sempéré et al. (1993), Rommevaux et al. (1994) and Gente et al. (in press).

The aim of this study is to investigate the observed variability of the axial domain, and to test the influence of a possible non steady-state thermal anomaly beneath the axis. Such a thermal anomaly closely resembles a narrow thermal intrusion at the axis as already introduced many years ago in thermal models like those of Sleep (1969, 1974 and 1975). But because of their steady-state description of the accretion, these as well as other models (Phipps Morgan and Forsyth, 1988; Buck and Su, 1989; Sotin and Parmentier, 1989; Parmentier and Phipps Morgan, 1990; Blackman and Forsyth, 1992; Cordery and Phipps Morgan, 1992) are based on an unequivocal relationship between spreading rate and vertical mantle flow at the axis. As shown by high-resolution surveys, this large-scale modelling of the accretion is inadequate within the axial domain itself. More specifically, it does not take into account the likely variability of the accretionary process through time. We address this question by adding to the spreading a second phenomenon, which we model by introducing a variable heat supply with time. In first instance, we refrain from any a priori assumptions on the relationship between spreading and heat supply.

Thermal modelling of accretion

Numerical simulation of accretion

In our model, accretion is viewed as the superposition of two processes, continuous seafloor spreading and variable thermal inputs associated with magmato-tectonic cycles. The model is illustrated in Fig. 1.

To simulate spreading, a new two-dimensional zone is created at the axis at regular time intervals (Fig. 1, upper part). The width of this zone is chosen to be 1 km, thus the spreading time interval dt depends on the spreading rate value: 50,000 yr for a full rate of 20 mm/yr, 5000 yr for 200 mm/yr. At each time step dt, a new axial zone is instantaneously emplaced, and this new axial zone moves the two half-domains aside. The newly-created zone has the same vertical temperature profile T_{axis} as the adjacent neighbouring zone at the given time such that no additional heat is introduced at the axis.

Superimposed on the spreading process, a magmato-tectonic cycle is simulated as a heat supply (temperature T_{hs}) occurring periodically in a 10 km wide axial conduit, going up to 4 km below the seafloor (Fig. 1, lower part). As hot material rises very rapidly to the surface (Dick, 1989), this thermal upwelling is assumed to be instantaneous. We have chosen the simple hypothesis of a periodic rather than random phenom-

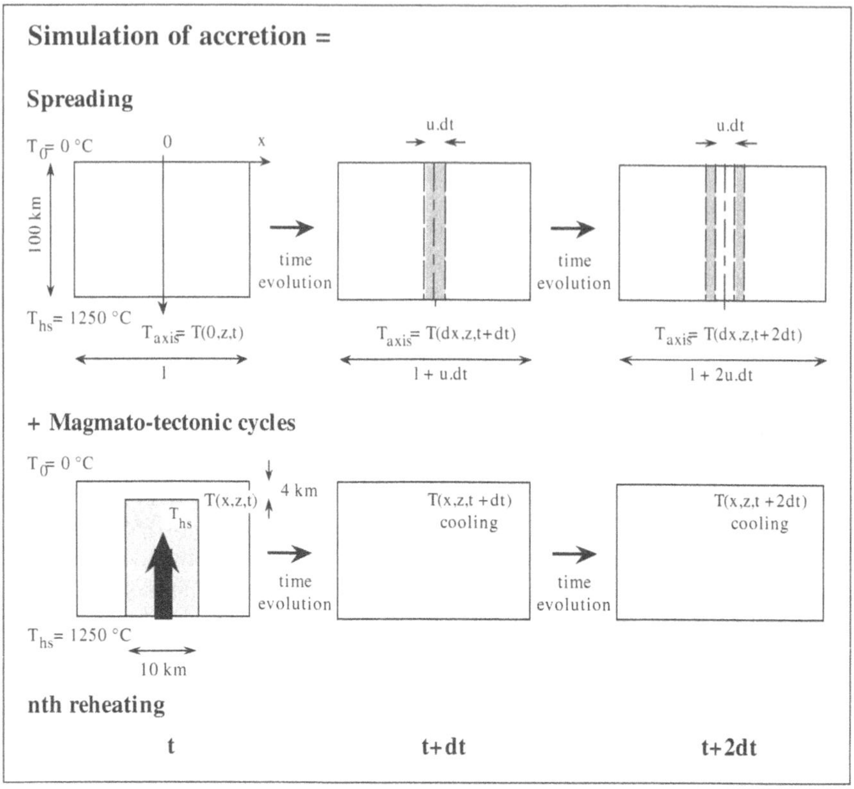

Figure 1. Diagram illustrating numerical model to simulate accretion as the superposition of seafloor spreading (upper part) and magmato-tectonic cycles (lower part).
The time evolution of the computational domain is shown from time t at the beginning of a cycle (left hand side) to time t+2dt, i.e., two spreading time steps afterward (right hand side).
Notations: x, across-axis distance; z, depth; t, time; u, full spreading rate; T, temperature; l, full x-dimension of the computational box at t; T_0 and T_{hs}, upper and lower boundary temperatures; T_{axis}, temperature at the ridge axis and in newly created zone.

enon, and the period in question is set equal to the duration of a magmato-tectonic cycle.

The present model simulates thermal evolutions characterized by a succession of cycles of reheating followed by cooling. The temperature distribution is computed numerically step by step in both space and time. Below we discuss the model predictions for two magmato-tectonic cycle durations (using maximum and mean values deduced from Table 4), for the cases of slow (20 mm/yr) and fast (200 mm/yr) spreading ridges. Table 5 summarizes the parameters of the four computational runs presented in this study. Only the time frequencies of the two principal phenomena (spreading and heat supply) differ from one run to another; all other physical parameters discussed below are set identical for slow and fast spreading ridges.

Equations and numerical technique

Equations

The basic equation in our thermal model is the two-dimensional (2D) heat equation for a non steady-state regime, taking into account horizontal and vertical conductive heat transfer, and heat loss due to partial melting:

$$\frac{\partial}{\partial t}T(x,z,t)=K\left(\frac{\partial^2}{\partial x^2}+\frac{\partial^2}{\partial z^2}\right)T(x,z,t)-\left(\frac{L}{c}\right)\frac{\partial}{\partial t}M(x,z,t)$$

where:

T(x,z,t)	temperature as a function of the across-axis distance x, depth z, and time t
K	thermal diffusivity
c	specific heat
L	latent heat of melting
M(x,z,t)	melt fraction depending on T(x,z,t)

In this model, the numerical solution is two-dimensional, although the thermal input simulating the magmato-tectonic cycles should likely be three-dimensional (3D), in particular for slow spreading ridges. A resolution in three dimensions needs much more complicated numerical developments, but it is clear that a 3D conduit would be required to further investigate the along-axis heat supply variations. Nevertheless, a 2D model is necessary first, as it allows us to test the possible range of the temporal variations of the thermal structure and melt fraction distribution.

The heat transfer associated with spreading, currently introduced as a convective term in the heat equation, is partly taken into account in our model via a succession of temperature reinitiations through time as outlined below. This way of introducing the spreading produces very small differences in the computed temperatures (<1%, except within the first 3 km beneath the upper boundary where a maximum error of 5% is reported) compared with other models such as the half-space model of lithospheric cooling (Davis and Lister, 1974).

No additional heat contribution is introduced in equation (1) because the radiogenic heat production in the oceanic crust is too low (Wakita et al., 1967). The thermal effect of hydrothermal circulation, often modelled as a heat loss at shallow depths (e.g., Morton and Sleep, 1985), is not taken into account in this study. There is insufficient information at present to constrain the temporal evolution of hydrothermal circulation, however, it is clearly episodic (Rona et al., 1983) such that a steady-state heat loss would be unrealistic. Moreover, the effect of hydrothermal circulation is probably limited to shallow levels and does not significantly affect the thermal state at greater depth (Morton and Sleep, 1985).

The melt fraction M(x,z,t) is expressed as a function of temperature and pressure, from McKenzie (1984):

$$M(x,z,t) = 100\,\frac{T(x,z,t) - T_{sol\,0} - S_{sol}\cdot p(x,z,t)}{T_{liq\,0} - T_{sol\,0} - \left(S_{sol} - S_{liq}\right)\cdot p(x,z,t)}$$

where:

$M(x,z,t)$	melt fraction (in percent)
$T(x,z,t)$	temperature as a function of the across-axis distance x, depth z, and time t
$p(x,z,t)$	pressure as a function of the across-axis distance x, depth z, and time t
$T_{sol\,0}$	solidus temperature at z = 0
$T_{liq\,0}$	liquidus temperature at z = 0
S_{sol}	slope of the solidus
S_{liq}	slope of the liquidus

Parameter values used in the computation are listed in Table 6.

For each computational step in time Δt (see below), we first solve equation (1) assuming that no melting occurs, and obtain the distribution of the mantle potential temperature. For regions in which this potential temperature is above the solidus, temperatures are calculated again, taking into

Table 6. Physical parameters

Variable	*Meaning*	*Value*	*Units*
Heat equation parameters			
ρm	density of mantle at 0°C	3.33	
ρ	density of mantle at temperature T $=\rho m(1-\alpha T)$		
α	thermal expansion coefficient	$3.28\ 10^{-5}$	$°C^{-1}$
k	thermal conductivity	3.14	$W\ m^{-1}\ °C^{-1}$
K	thermal diffusivity	$1.13\ 10^{-6}$	$m^2\ s^{-1}$
c	specific heat=$k/\rho K$		
L	latent heat of melting	720	$J\ kg^{-1}\ °C^{-1}$
Melting equation parameters			
T_{sol0}	solidus temperature at z = 0	1115	°C
T_{liq0}	liquidus temperature at z = 0	1715	°C
S_{sol}	slope of the solidus	120	$°C\ GPa^{-1}$
S_{liq}	slope of the liquidus	-16	$°C\ GPa^{-1}$
Boundary conditions			
T_0	upper boundary temperature	0	°C
T_{hs}	lower boundary temperature	1250	°C

account the latent heat of melting. Finally, the melt fraction distribution is deduced from equation (2).

Experimental results on non steady-state partial melting relations are currently not available. Both latent heat of melting and melting relationships evolve with time. The observed compositions of basalts erupted along a ridge axis are generally uniform, suggesting that melting is an isochemical process. We have assumed that the latent heat of melting remains constant wherever melting occurs, ignoring the fact that it depends slightly on pressure as well as on melt composition. Solidus and liquidus curves probably evolve with time toward higher temperatures, in relation with the previous melting history of the material (Jaques and Green, 1980). Unfortunately, there are no fractional melting laws available which would allow us to quantify such a process (McKenzie, personal communication).

We wish to emphasize that all of the factors mentioned above (i.e., introduction of the third dimension, variable latent heat of melting, evolution of the melting relationship with time, thermal effect of the hydrothermal circulation) will only enhance the strong trend due to the occurrence of cooling periods through time. This means that our results represent an upper (hotter) limit for the thermal regime beneath spreading centres.

Spatial and temporal discretization

The solution of the differential equation (1) is obtained by the finite difference method. The numerical integration is based on alternating-direction implicit methods (Douglas, 1955). For this study, the computational steps are as follows:

in space,	$\Delta x = \Delta z = 500$ m	
in time,	$\Delta t = 5000$ yr	for a slow-spreading ridge
	$\Delta t = 500$ yr	for a fast-spreading ridge

The value of the spatial step Δx has been chosen to have a reasonable number of computational nodes within both the new axial zone created by spreading (here 1 km wide, i.e., 3 nodes) and the axial thermal conduit (here 10 km wide, i.e., 21 nodes). The dimensions of the computational domain are 100 km in depth, and 100 km in the x-direction at time t=0 when initial conditions are set. This last x-dimension (shown as l in Fig. 1) increases with time as the result of spreading. A good convergence of the discrete solution in time needs a number of iterations of about 20 for such a grid (Peaceman and Rachford, 1955), i.e., a time interval of 10 times Δt between two successive temperature results.

Boundary conditions

The temperature is set to a constant both at the upper boundary (equal to T_0) and at the lower boundary (equal to T_{hs}, see Table 6 for numerical val-

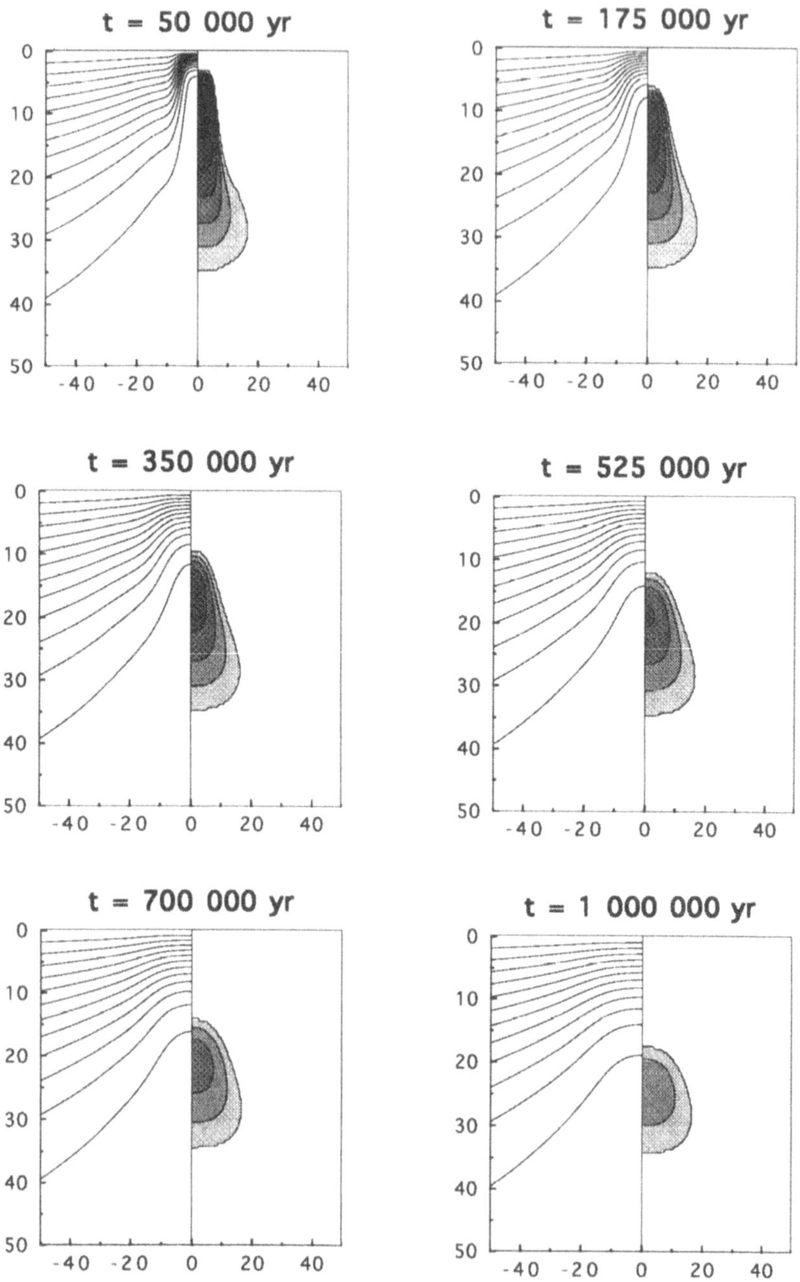

Figure 2. Modelling results showing evolution with time of the thermal structure and melting zone for a slow spreading ridge with a full spreading rate (u) of 20 mm/yr and a cycle duration of 1 Ma (computational run 1). Plots for elapsed time as indicated, from 50,000 yr to 1,000,000 yr since start of cycle. Domains shown (100 km x 50 km) are divided in halves down the ridge axis, with the thermal structure to the left, shown in isotherms ranging from 0°C to 1200°C in steps of 100°C, and melt fraction to the right, shown in shades of grey ranging from 0% (white) to 21% (black) in steps of 3%.

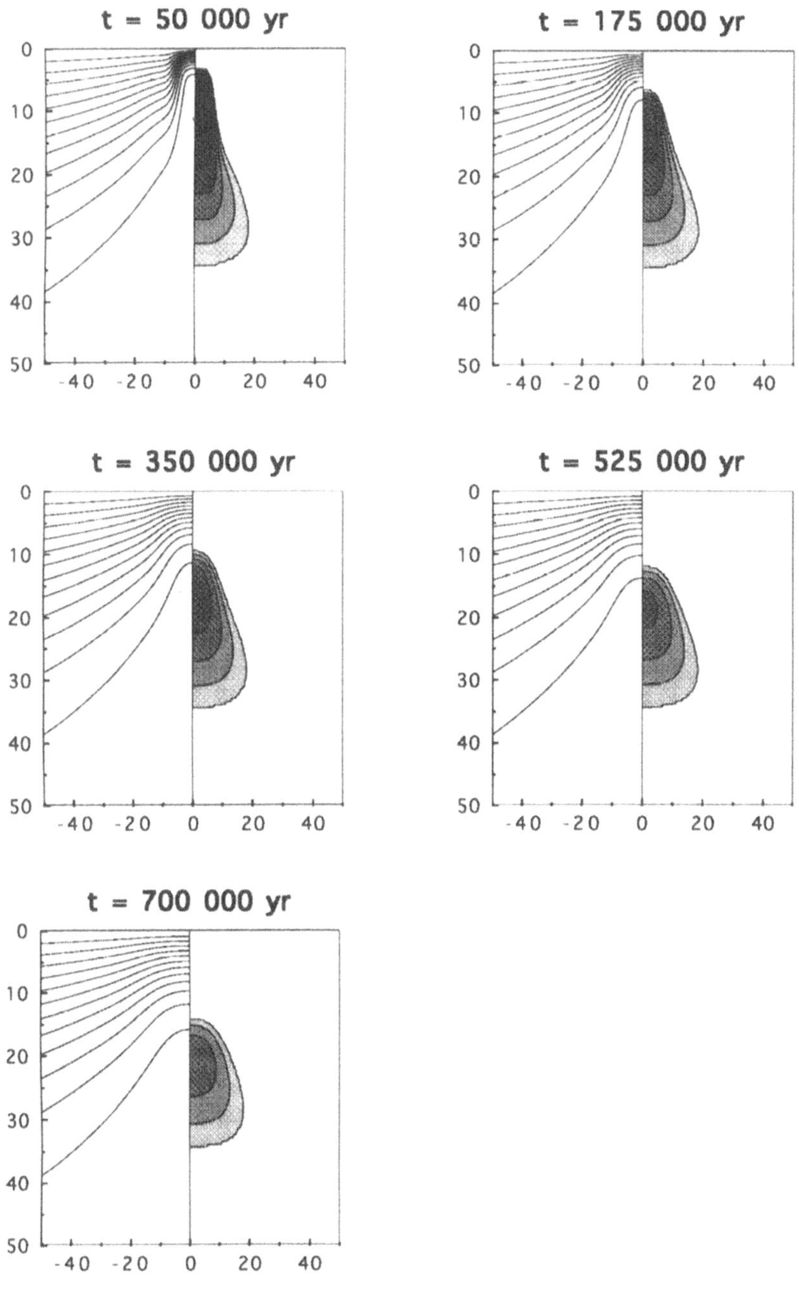

Figure 3. Modelling results showing evolution with time of the thermal structure and melting zone for a slow spreading ridge with a full spreading rate (u) of 20 mm/yr and a cycle duration of 700,000 yr (computational run 2). Plots for elapsed time as indicated, from 50,000 yr to 700,000 yr since start of cycle. Legend as in Figure 2.

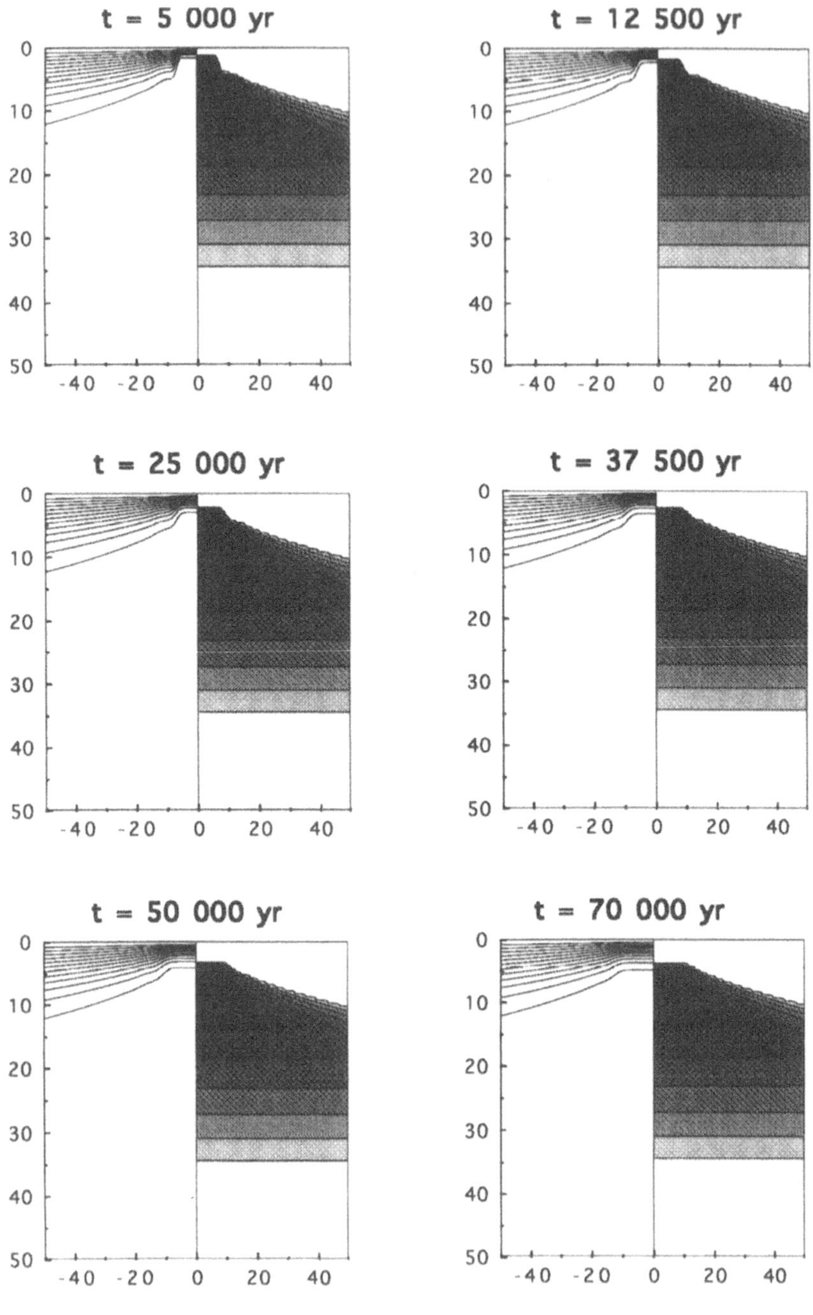

Figure 4. Modelling results showing evolution with time of the thermal structure and melting zone for a fast spreading ridge with a full spreading rate of 200 mm/yr and a cycle duration of 70,000 yr (computational run 3). Plots for elapsed time as indicated, from 5,000 yr to 70,000 yr since start of cycle. Legend as in Figure 2.

ues). At lateral boundaries, the value of lateral heat flow is set to zero. In order to avoid edge effects in the zone of interest, these lateral boundaries are chosen sufficiently far away from this zone.

Initial conditions and reinitiations at regular time steps
At time t=0, the initial temperature distribution is calculated from the half-space model of lithospheric cooling (Davis and Lister, 1974). Near the ridge, this model is equivalent to the plate model of McKenzie (1967) or Parsons and Sclater (1977).
At regular time steps, we reinitiate the temperature distribution in the axial domain, either in a new narrow zone created by the spreading, with each spreading time step dt (Fig. 1, upper part), or in the axial conduit at the start of each magmato-tectonic cycle (Fig. 1, lower part), or both. Table 5 gives time steps of both temperature reinitiation processes used for the four computational runs presented in this study.
The thermal evolution is computed through time for computational time steps Δt given previously (one order of magnitude smaller than the two time steps of Table 5). We only use the temperature results obtained after 5 or 6 magmato-tectonic cycles, when the thermal regime is stabilized for 2 successive cycles.

Modelling results: thermal structure and melt fraction

Below we first present modelling results directly deduced from equations (1) and (2) describing the evolution with time of the thermal structure and melt fraction. These results are summarized in Figs. 2 and 3 for slow spreading ridges, and in Fig. 4 for fast spreading ridges, showing the evolution through one cycle. The results of the fourth computational run listed in Table 5 do not significantly differ from those in Fig. 4 (5 first plots) and are not shown. Implications of the modelling results for crustal production are discussed in a next section.

Ridge thermal structure

At a slow spreading ridge, the axial domain represents a highly variable thermal state, evolving periodically from a hot state toward cooler ones as each cycle proceeds (Figs. 2 and 3). At the axis, prevailing temperatures can decrease, through a cycle, to 50% of their initial values at 5 km depth and to 25% at 10 km depth, but temperatures remain constant at depths greater than 20 km. The axial conduit shape vanishes quickly with progressive cooling. For the longest cycle of run 1, the overall thermal regime is somewhat colder than for run 2, with a cooler stage in particular at the end of the cycle. Off-axis, for distances greater than 20 km away from the axis, there is no significant variation with time. At less than 50 km from the axis, the temperature does not vary for depths greater than 40 km even though there is a boundary temperature condition at a depth of 100 km.

In contrast, for a fast spreading ridge, the axial domain always remains in a hot thermal state during the entire cycle (Fig. 4). For the same time t after the beginning of a cycle, no difference is recorded between run 3 (cycle of 70,000 yr) and run 4 (cycle of 50,000 yr). Fast spreading has a strong effect on the shape of the isotherms, drawing the trail of the initial conduit away from the axis. The axial domain temperature varies with depth only in the first 10 km. In fact, our calculated thermal structure is nearly steady-state in this case and, except in the axial domain itself, very close to the one deduced from classical thermal models such as Parsons and Sclater (1977). This agrees with depth observations near the axes of fast ridges, which generally differ very little from the subsidence curve of these classical models.

Following Nicolas et al. (1980), we may choose the 1000°C isotherm as the lithosphere/asthenosphere interface. For slow spreading ridges, the thermal lithosphere produced at the ridge axis would thicken from nearly 2 km for the hottest state to about 10 km at the end of a cycle immediately before the next reheating. For fast spreading ridges, the modelled lithospheric thickness oscillates little with time between 1 and 3 km at the ridge axis. At a distance of 20 km away from the axis, the lithosphere of slow ridges is 16 km thick, whilst at 50 km away from the axis it is 24 km thick without significant variation through time. At these same distances away from the axis of fast spreading ridges we predict values of 4.5 and 7 km, respectively.

Evaluation of the model

In order to evaluate our thermal model, we have compared its output with various types of parameters which are believed to be temperature-dependant. The brittle/ductile transition, taken as the 800°C isotherm (Tapponnier and Francheteau, 1978; Harper, 1985), may be one of these parameters. From the thermal structures computed for slow spreading ridges we infer that its depth varies greatly at the axis, between 2 and 8 km (Figs. 2 and 3). This range of values is in good agreement with the seismicity observed beneath slow ridges. Beneath the deepest and largest axial valleys, presumed by many authors (for example Harper, 1985; Sempéré et al., 1993) to correspond to the coldest thermal structures, seismic activity extends to the greatest depths (often 6-8 km and up to a maximum of 8.8 km), whereas lesser depths are observed for seismic activity underneath neighboring narrower valleys and relative topographic highs inside the same ridge segment (Murray et al., 1984; Toomey et al., 1985 and 1988; Huang et al., 1986; Huang and Solomon, 1988; Kong et al., 1992; Barclay et al., 1993; Wolfe et al., 1993). The large variable fault scarps, revealing a high roughness which characterizes slow ridge bathymetry (Menard, 1967;

Macdonald, 1982; Malinverno and Pockalny, 1990; Malinverno, 1991), also require the existence of a thick brittle layer. In the case of fast spreading ridges, the brittle layer thickness deduced from our model is small and shows little variation, from 0.8 to 1.8 km (Fig. 4), in agreement with the smooth bathymetry observed (Menard, 1967; Macdonald, 1982; Malinverno and Pockalny, 1990; Malinverno, 1991). Very few earthquakes are recorded beneath fast ridge axes, and only at shallow depths (Riedesel et al., 1982; Tréhu and Solomon, 1983; Wilcock et al., 1992).

Another possible test of the model is the thickness of the magnetized layer, which can be viewed as the depth of the Curie's isotherm of about 600°C (580°C for gabbros). In the case of slow spreading ridges, our model gives axial thicknesses for this layer ranging from 0.9 km to a maximum value of 5.5 km. For the M.A.R. near 26°S, Grindlay et al. (1992) infer a thickness of the magnetized layer varying along a segment between 0.8 km for a hot zone at the centre and 2.6 km at the ends. Little information is available about the thickness of the magnetized layer produced at fast spreading ridges, but our model implies little variation between 0.6 km and 1.4 km. Seismic tomography yields a qualitative constraint, revealing the presence of hot material at shallow depths (2-3 km) along the whole segment of a fast ridge (e.g., Toomey et al., 1990), which places an upper limit to the magnetic layer thickness.

Another way to evaluate our model involves computation of the effects on gravity anomaly patterns and thermal subsidence with time, and to compare the model predictions with mantle Bouguer anomalies and morphology observed on spreading ridges. A study of this type is currently in progress (Tisseau and Tonnerre, in preparation).

The agreement between the calculated thermal structures and some known constraints on the ridge axis domain is encouraging, and suggests that the models presented above predict realistic thermal regimes beneath ridge axes. As an additional control we have compared the present model results with results from previous models with broadly similar assumptions. For example, the steady-state thermal structure presented by Sleep (1974) for a slow spreading ridge falls within the range of our evolution through time (Fig. 2), and may be viewed as a time-averaged structure. Sleep (1974) has already noted that at slow spreading rates a thick cool region exists at the axis, whereas at high spreading rates hot mantle material may intrude till almost near the surface. Our model allows us to quantify the fluctuations with time around such an average thermal state.

Partial melting

Before presenting our results on partial melting, we must discuss the influence of one parameter, i.e., the heat supply temperature T_{hs}, which is set here equal to the temperature of the lower boundary of the computational domain. The highest melt fraction occurs at shallow depths and at the beginning of a cycle, and its value depends on the temperature gap between the intrusion temperature T_{hs} and the solidus temperature near the surface (i.e., $T_{sol\,0}$). We have investigated the effect of varying T_{hs} (from 1300°C to 1200°C) using the partial melting relations described above for constant parameter values as listed in Table 6. The highest computed melt fractions are, for example, 28% for T_{hs}=1300°C, 19% for 1250°C, and 10% for 1200°C in the case of slow spreading ridges, and 30%, 21%, and 13%, respectively, for these same T_{hs} values in the case of fast spreading ridges. We prefer the 1250°C value, because it yields a likely realistic highest melt fraction of about 20%, compatible with observations on both slow and fast spreading ridges.

For a slow spreading ridge, the melting zone deduced from our model is pear-shaped at the beginning of a cycle and evolves, mainly from the top (Figs. 2 and 3), towards a more circular feature during progressive cooling. Its maximum width is about 30 km. Its upper limit gradually moves downward from about 3 to about 17 km below seafloor, whilst its lower limit remains at a nearly constant depth of about 30 km. These results discard the idea of a wide region of melting beneath a slow spreading centre, but approximate models involving a narrow tall diapir beneath the axis (Sleep, 1974; Cordery and Phipps Morgan, 1992), but this diapir waxes and wanes through time. The melting region is more confined in run 1 than in run 2, for a longer duration of cycles, i.e., for an overall colder thermal regime. The highest melt fractions occur at the beginning of a cycle. The maximum value for run 1 decreases from 19% to 5.3% at the end of the 1 Ma cycle, whilst for run 2 it decreases from 19% to 8.2% at the end of the 700,000 yr cycle (compared with 7.6% at time 700,000 yr of run 1, Fig. 5). The variability in the melt fraction deduced from our model and its range of values are consistent with the large variability in the type of mantle rocks observed along slow ridge segments away from any hotspot influence: most deep rocks reported inside ridge segments away from discontinuities are

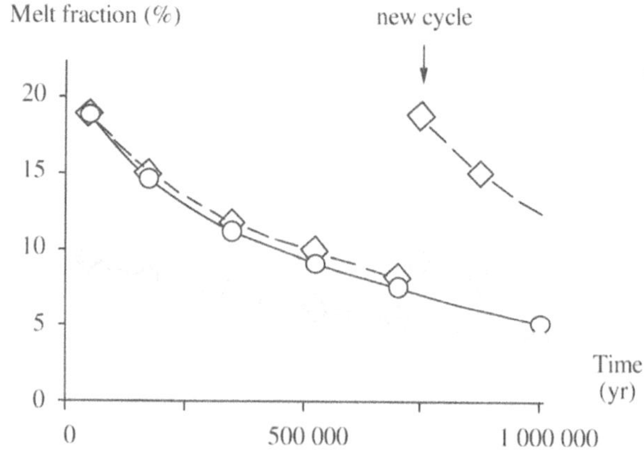

Figure 5. Computed melt fractions versus time for a slow spreading ridge, full spreading rate 20 mm/yr. Maximun melt fractions shown in black, average melt fractions shown in grey. Solid lines with open circles represent run 1 with a cycle duration of 1 Ma. Dashed lines with diamonds denote run 2 with a cycle of 700,000 yr. Run 2 involves new cycle starting at 700,000 yr.

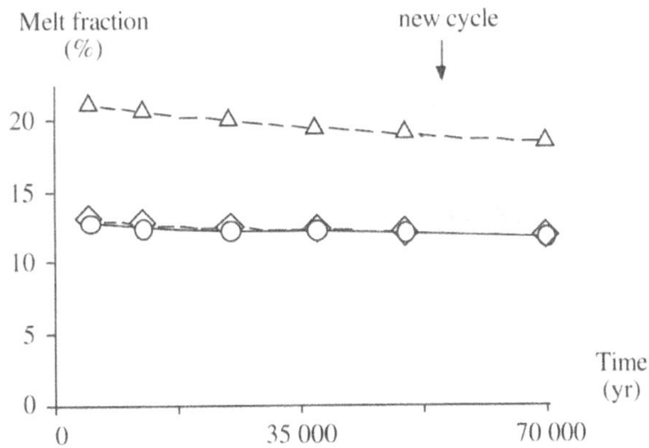

Figure 6. Computed melt fractions versus time for a fast spreading ridge, full spreading rate 200 mm/yr. Black lines represent run 3 (cycle of 70,000 yr), grey lines run 4 (cycle of 50,000 yr). Dashed lines with triangles denote maximum melt fractions. Average melt fractions have been computed for the case of a crustal production zone of 15 km (dashed lines with diamonds) and 20 km wide (solid lines with open circles). Between times 0 and 50,000 yr, results of run 4 are identical to (and undistinguishable from) those of run 3. Run 4 involves new cycle starting at 50,000 yr.

primitive products of minor partial melting, although limited occurrences of depleted rocks do exist (see references in Table 2).

In the case of fast spreading ridges much hotter than the slow ones, the melting region has a very different triangular shape, with a large lateral extent beneath the spreading centre (Fig. 4). Its upper limit remains very close to the surface, at shallow depths varying between 1 and 4 km. The depth of its lower limit is constant through time and equals about 34 km, a value close to that for slow ridges. The maximum melt fraction varies little with time, and decreases from 21.2% at the beginning of a cycle to 18.5% at the end of the longest cycle (Fig. 6). The highest value (19%) obtained in the case of slow ridges only occurs at the beginning of cycles and falls within this range. The region of highest melt fraction (>18%) diminishes with time, mainly through deepening of the upper boundary of the melting zone from 2 down to 6 km (Fig. 4). All these results agree well with the marked homogeneity reported from the few dredged and drilled mantle rocks along fast ridge axes, all showing high degrees of melting (see references in Table 2 for Hess Deep).

We may also compare our model predictions with data on ophiolitic peridotites. Studies in ophiolites (see review by Nicolas, 1989) reveal the existence of two main ophiolite types: the Harzburgite Ophiolite Type with a depleted mantle and a thick crust which would correspond to fast spreading rates, and the Lherzolite Ophiolite Type with a non-depleted mantle and a thin crust which would correspond to slow spreading rates.

Based on the same principles for all ridges, our model reconciles opposing views on the shape and extent of the melt region beneath spreading centres: either a very wide region varying little with time in the case of a very hot ridge, or a narrow region waxing and waning through time in the case of a ridge with a fluctuating thermal state. Key factors are the time frequencies of the different accretionary processes (spreading and heat supply), because these have a major influence on the thermal regime beneath the ridge. The lower boundary of the melt region lies at a nearly constant depth of the order of 30-34 km. The exact value in the model is linked to the choice of both melting relationships and lower boundary temperature T_{hs}. Nevertheless, in contrast with other models (Phipps Morgan and Forsyth, 1988; Sotin and Parmentier, 1989; Buck and Su, 1989; Parmentier and Phipps Morgan, 1990), our model predicts that this lower boundary cannot be very deep.

Discussion: implications for crustal production

Assumptions

On the basis of the thermal structures and melt fractions calculated from our model, we now explore the implications for crustal production at ridge axes. Several assumptions are required. First, an instantaneous extraction of melt is considered in the computations because of the extremely high melt velocity (see review in Nicolas, 1989) relative to accretion time scales. The second assumption concerns the zone of extraction. Two main alternative models have been proposed in the literature for melt extraction, i.e., melt extraction by hydrofracturing (e.g., Nicolas, 1989), and by porous flow (e.g., McKenzie, 1984). We have used a porous-media flow mechanism because hydrofracturing, a concept essentially proposed on the basis of ophiolite data, is probably more important in shallow rather than deep processes. Another criterion in favor of choosing a porous media flow model for our computations is the smaller threshold value for extracting melt, i.e., a melt fraction of 3%, rather than the 7% in the case of extraction by hydrofracturing. Choosing the higher threshold of 7% will only emphasize the results obtained with the 3% threshold, and lead to an increase of, in particular, the time spans without any crustal production.

Figures 2, 3 and 4 show the time-dependent spatial extent of partially molten zones, contoured for melt fraction values between 0% and 21% in steps of 3%. Using a 3% threshold, the 3% contour also delineates the potential zone of melt extraction. Though somewhat smaller, this zone has a similar shape as the zone where first melting occurs. Its lower boundary, at a depth of about 30 km, depends on the parameter values used, in particular on the temperature of first melting. This would allow a comparison with the source depth of basalts provided that there would be ways to independently assess the melting laws in question. Inside the potential zones of melt extraction as predicted by the model (i.e., the domains with calculated melt fractions higher than 3%), an average melt fraction is calculated and compared with the melt fraction of basalts: the common assumption of mixing at depth is made here.

Figure 7. Computed crustal production versus time for a slow spreading ridge, full spreading rate 20 mm/yr. Solid line with open circles from run 1 (cycle of 1 Ma), dashed line with diamonds from run 2 (cycle of 700,000 yr). Run 2 involves new cycle starting at 700,000 yr.

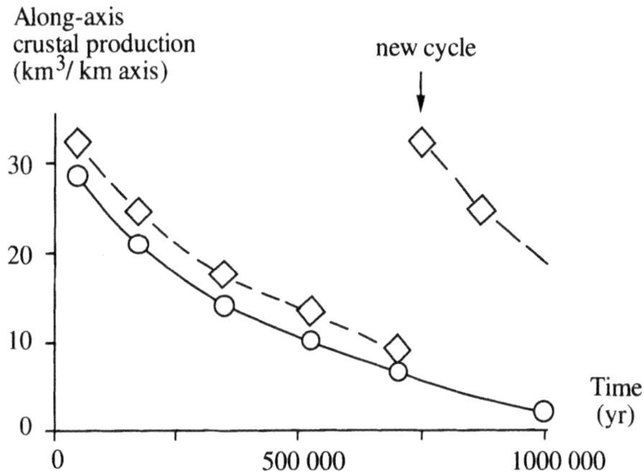

On the basis of the simple assumption that the integrated melt production equals the amount of melt extraction, some estimates can be made on crustal production. These estimates represent upper bound values for which all extractable melt is in fact extracted.

Model predictions and comparison with data

Slow spreading ridges

For slow spreading ridges, the average value of melt fraction decreases with time from 9.5% at the beginning of a cycle to 4% at the end (Fig. 5). There are no significant differences between runs 1 and 2, except in the smaller values towards the end of a cycle. The predicted crustal production shows large variations at slow ridge axes (Fig. 7), and decreases strongly with time from a maximum production of the order of 30 km^3 (per km length of ridge axis). For a sufficiently long cycle, this production decreases toward zero. Such a variability in the production yields a theoretical crustal thickness ranging between 6.4 and 0.5 km if distributed on an across-axis domain of 5 km wide, and between 3.2 and 0.2 km on a 10 km wide domain. On slow spreading ridges, volcanism seems to be rather confined in a narrow domain, and 5 km would be a more realistic maximum value for such a distribution. This means that a 6 km thick normal crust could be present at the axis at the beginning of a cycle when the underlying mantle is hottest. As cycles become longer, the ridge axis is colder and there may be no crustal production during significant periods of time. Mantle rocks will then be emplaced along the axis in the coldest regions, in first instance at the ends of a segment where their abundance is greatest, but also within a segment where their abundance is related to the stage of advanced cooling of the axial domain. On the other hand, towards the ends of cycles, the thicker lithosphere would also favor cooling of rising magma

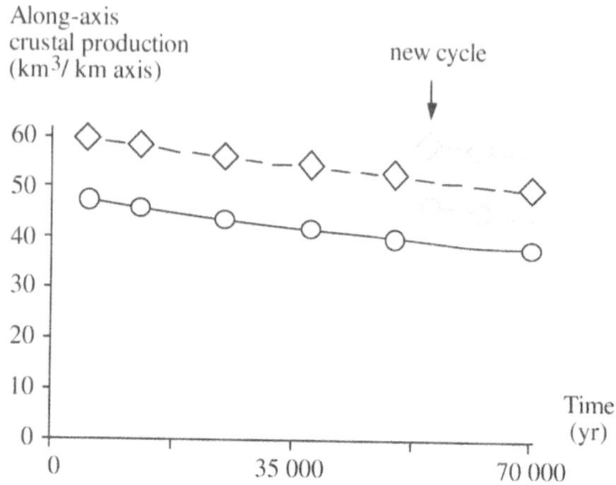

Figure 8. Computed crustal production versus time for a fast spreading ridge, full spreading rate 200 mm/yr. Black lines from run 3 (cycle of 70,000 yr), grey lines from run 4 (cycle of 50,000 yr). Dashed lines with diamonds for a crustal production zone of 20 km; solid lines with stars for a 15 km wide crustal production zone. Results of run 4 are identical to (and undistinguishable from) those of run 3, for all times between 0 and 50,000 yr. Run 4 involves new cycle starting at 50,000 yr.

at depth, leading to the emplacement of gabbroic bodies as in the model proposed by Lagabrielle and Cannat (1990) or Cannat (1993).

Fast spreading ridges

The case of fast spreading ridges is very different. The potential zone of melt extraction is very large, and laterally unlimited in our model (Fig. 4). This agrees with the observation that volcanism can be important away from ridge axes (e.g., Batiza et al., 1990; Carbotte and Macdonald, 1992). Restricting the zone of melt extraction to a width of 20 km, similar to the case of slow spreading ridges (Figs. 2 or 3), the average melt fraction varies very little, i.e., from 12.7% to 11.8%. For a 15 km wide zone of melt extraction, these values range between 13% and 11.8% (Fig. 6).

A theoretical crustal production may be calculated in a similar manner as for slow spreading ridges. Its value varies little with time and remains always high between 59 and 51 km^3 (per km length of ridge axis) for a 20 km wide extraction zone, and between 47 and 39 km^3 for a 15 km wide region (Fig. 8). If this production is distributed at the surface over an across-axis domain of 5 km wide, the evolved crustal thickness should be 10-12 km for a melt extraction zone of 20 km, and 8-10 km for a melt extraction zone of 15 km. Distributed over a 10 km wide region, these values become 5-6 km for the first, and 4-5 km for the second case. These values are close to the seismic crustal thicknesses of fast spreading ridges. The crustal thickness and its variations allow us to constrain the width of the melt extraction zone. Spiegelman and McKenzie (1987) report larger widths, from 37 to 60 km, depending on the viscosity. The first value seems more in agreement with our results, but further constraints are needed on both fractional melting relations and mantle temperature at depth to reconcile such discrepancies.

In contrast to slow spreading ridges, the calculated crustal production for fast spreading ridges is high and nearly constant with time, which is in good agreement with the homogeneous thick crust observed along the whole of fast spreading ridge axes. No mantle rock outcrop is then likely to occur under such conditions. On the other hand, the calculated average melt fraction is higher for fast than for slow spreading ridges, which concurs with the presence of more primitive basalts in the case of slow than in the case of fast spreading ridges (Niu and Batiza, 1993). These authors emphasize two along-axis geochemical trends in the compositions of basalts: a local trend, characteristic of slow spreading ridges and involving the lack of interconnection between melt diapirs along the axis, as opposed to a global trend indicating a high degree of melt mixing seen along fast spreading ridge axes. This idea is supported by the very different length scales of our modelled extracting zones.

Summary and conclusions

We have designed a thermal model which accounts for two first-order processes of accretion, i.e., seafloor spreading and thermal inputs associated with periodic magmato-tectonic cycles. Such a quantification of the thermal regime beneath ridge axes and the inherent implications for partial melting and crustal production lead to the following conclusions:

(1) The time frequencies of seafloor spreading and thermal inputs associated with magmato-tectonic cycles essentially control how any given ridge classifies between two end-member type ridges: a nearly steady-state, very hot ridge and a "fluctuating" ridge where short periods of heat input are separated by longer periods of cooling.

(2) On the basis of the present knowledge of ridges, different realistic time frequencies have been tested in the model. For high frequencies, the axial domain has no time to cool significantly between two successive reheatings, and its thermal state approximates that of existing steady-state models. In contrast, lower frequencies allow sufficient cooling, which has a drastic effect on the thermal state of the axial domain, hence on thermally induced features like melt fraction and crustal production, most notably in the case of slow spreading ridges.

(3) The partial melting zones computed in this model study range between two end-members. This reconciles opposing views regarding the melting regions beneath spreading centres: a wide steady-state region for hot ridges as opposed to a narrow tall diapir for a "fluctuating" ridge. This diapir is not a steady-state phenomenon but waxes and wanes, which accounts for the large variability, most notably in the types of mantle rock, observed both along and across ridge axes.

(4) The present numerical model for oceanic ridges incorporating a fluc-

tuating heat supply predicts that mantle rocks will be emplaced at the axis only if sufficient cooling occurs. The exposure of upper mantle peridotites is thus more likely in case of magmato-tectonic cycles of long duration, and reveals the coldest underlying mantle domains along the axis.

Acknowledgements

We would like to thank all researchers from the group DORSALES of Brest, and also M. Cannat, J. Girardeau, D. McKenzie and J. Phipps Morgan, for fruitful discussions about this paper. Particular thanks are due to J. Francheteau, J. Goslin, J. Karsten, R. Vissers and two anonymous reviewers for many helpful suggested improvements in the text. One of us (T.T.) has a doctoral grant from the French Ministry of Research and Technology. This work was partially funded by CNRS-Institut National des Sciences de l'Univers DBT grant 92.38.23 Thème Instabilités.

References

Arai, S. and Fujii, T., 1979. Petrology of ultramafic rocks from site 395. Initial Rep. Deep Sea Drill. Proj., 45: 587-594.

Auzende, J.-M., Eissen, J.-P., Lafoy, Y., Gente, P. and Charlou, J.-L., 1988. Seafloor spreading in the North Fiji Basin (Southwest Pacific). Tectonophysics, 146: 317-351.

Auzende, J.-M., Gracia-Mont, E., Bendel, V., Lafoy, Y., Lagabrielle, Y., Okuda, Y. and Ruellan, E., (in press), Morphologic variations at an intermediate rate spreading ridge (North Fiji Basin). In: Wiley etal. (Eds), Mantle and crustal process in Mid-Ocean Ridges. in press.

Barclay, A.H., Toomey, D.R., Purdy, G.M. and Solomon, S.C., 1993. FARA microearthquake experiments III: results from the Mid-Atlantic Ridge at 35°N. Eos, Trans. AGU, 74: 601.

Bassias, Y. and Triboulet, C., 1992. Petrology and P-T-t evolution of the South West Indian Ridge peridotites. A case study: East of the Melville Fracture Zone at 62°E. Lithos, 28: 1-19.

Batiza, R., Melson, V. and O'Hearn, T., 1988. Simple magma supply geometry inferred beneath a segment of the Mid-Atlantic Ridge. Nature, 335: 428-431.

Batiza, R., Niu, Y. and Zayac, W.C., 1990. Chemistry of seamounts near the East Pacific Rise: implications for the geometry of sub-axial mantle flow. Geology, 18: 1122-1125.

Bell, R.E. and Buck, W.R., 1992. Crustal control of ridge segmentation inferred from observations of theReykjanes Ridge. Nature, 357: 583-586.

Bideau, D., Hébert, R., Hékinian, R. and Cannat, M., 1991. Metamorphism of deep-seated rocks from the Garrett Ultrafast Transform (East Pacific Rise near 13°25'S). J. Geophys. Res., 96: 10079-10099.

Blackman, D.K. and Forsyth, D.W., 1991. Isostatic compensation of tectonic features of the Mid-Atlantic Ridge: 25-27°30'S. J. Geophys. Res., 96: 11741-11758.

Blackman, D.K. and Forsyth, D.W., 1992. The effects of plate thickening on three-dimensional, passive flow of the mantle beneath mid-ocean Ridges. In: J. Phipps Morgan, D.K. Blackman and J.M. Sinton (Eds), Mantle flow and melt generation at mid-ocean ridges. American Geophyssical Union, Washington, Geophysical Monograph 71, pp. 311-326.

Bonatti, E., Honnorez, J. and Ferrara, G., 1971. Peridotite-gabbro-basalt complex from the equatorial Mid-Atlantic Ridge. Phil. Trans. R. Soc. London., 268: 385-402.

Bonatti, E., Honnorez, J., Kirst, P. and Radicati, F., 1975. Metagabbros from the Mid-Atlantic Ridge at 6°N: contact-hydrothermal-dynamic metamorphism beneath the axial valley. J. Geol., 83: 61-78.

Bonatti, E. and Honnorez, J., 1976. Sections of the Earth's crust in the Equatorial Atlantic. J. Geophys. Res., 81: 4104-4116.

Bonatti, E., Craig Simmons, E., Breger, D., Hamlyn, P.R. and Lawrence, J., 1983. Ultramafic rock/seawa ter interaction in the oceanic crust: Mg-silicate (sepiolite) deposit from the Indian Ocean floor. Earth Planet. Sci. Lett., 62: 229-238.

Boudier, F., 1979. Microstructural study of three peridotite samples drilled at the western margin of the Mid-Atlantic Ridge. Initial Rep. Deep Sea Drill. Proj., 45: 603-608.

Buck, W.R. and Su, W., 1989. Focused mantle upwelling below mid-ocean ridges due to feedback between viscosity and melting. Geophys. Res. Lett., 16: 641-644.

Cannat, M., Bideau, D. and Hébert, R., 1990a. Plastic deformation and magmatic impregnation in serpen tinized ultramafic rocks from the Garrett Transform Fault (East Pacific Rise). Earth Planet. Sci. Lett., 101: 216-232.

Cannat, M., Juteau, T. and Berger, E., 1990b. Petrostructural analysis of the leg 109 serpentinized peri dotites. Proc. Ocean Drill. Prog., Init. Repts., 106/109: 47-56.

Cannat, M., Bideau, D. and Bougault, H., 1992. Serpentinized peridotites and gabbros in the Mid-Atlantic Ridge axial valley at 15°37'N and 16°52'N. Earth Planet. Sci. Lett., 109: 87-106.

Cannat, M., 1993. Emplacement of mantle rocks in the seafloor at mid-ocean ridges. J. Geophys. Res., 98: 4163-4172.

Carbotte, S. and Macdonald, K., 1992. East Pacific Rise 8°-10°30'N: evolution of ridge segments and dis continuities from SeaMARC II and three-dimensional magnetic studies. J. Geophys. Res., 97: 6959-6982.

CAYTROUGH, 1979. Geological and geophysical investigation of the Mid-Cayman Rise spreading center:initial results and observations. In: M. Talwani, C.G. Harrison and D.E. Hayes (Eds), Deep Drilling Results in the Atlantic Ocean: Ocean crust, Maurice Ewing Ser. American Geophysical Union, Washington, pp. 66-95.

Ceuleneer, G., Nicolas, A. and Boudier, F., 1988. Mantle flow patterns at an oceanic spreading centre: the Oman peridotites record. Tectonophysics, 151: 1-26.

Chadwick, W.W., Embley, R.W. and Fox, C.G., 1991. Evidence for volcanic eruption on the Southern Juan de Fuca Ridge between 1981 and 1987. Nature, 350: 416-148.

Constantin, M., Hékinian, R., Ackermand, D., Stoffers, P. and Francheteau, J., 1993. Upper mantle and lower crust exposed in the Easter microplate (South East Pacific). Terra Abstracts, 5: 184-185.

Cordery, M.J. and Phipps Morgan, J., 1992. Melting and mantle flow beneath a mid-ocean spreading cen ter. Earth Planet. Sci. Lett., 111: 493-516.

Courtillot, V., Achache, J., Landre, F., Bonhommet, N., Montigny, R. and Féraud, G., 1984. Episodic spreading and rift propagation: new paleomagnetic and geochronologic data from the Afar nascent passive margin. J. Geophys. Res., 89: 3315-3333.

Crane, K. and Ballard, R.D., 1981. Volcanics and structure of the FAMOUS narrowgate rift: evidence for cyclic evolution: AMAR 1. J. Geophys. Res., 86: 5112-5124.

Davis, E.E. and Lister, C.R.B., 1974. Fundamentals of ridge crest topography. Earth Planet. Sci. Lett., 21: 405-413.

Davis, E.E., Currie, R. and Sawyer, B., 1987. Bathymetry map 6- 1987, northern Juan de Fuca Ridge., Geological Survey of Canada, B. C., Sidney.

Dick, H.J.B., Thompson, G. and Bryan, W.B., 1981. Low-angle faulting and steady-state emplacement of plutonic rocks at ridge-transform intersection. Eos, Trans. AGU, 62: 406.

Dick, H.J.B., 1989. Abyssal peridotites, very slow spreading ridges and ocean ridge magmatism. In: A.D. Saunders and M.J. Norry (Eds), Magmatism in the ocean basins. Geological Society Special Publication, pp. 71-105.

Douglas, J.J., 1955. On the numerical integration of d2u/dx2 + d2u/dy2 = du/dt by implicit methods. J. Soc. Indus. Appl. Math., 3: 42-65.

Engel, C.G. and Fisher, R.L., 1975. Granitic to ultramafic rock complexes of the Indian Ocean Ridge system, western Indian Ocean. Geol. Soc. Am. Bull., 86: 1553-1578.

Forsyth, D.W., Ehrenbard, R.L. and Chapin, S., 1987. Anomalous upper mantle beneath the Australian-Antartic discordance. Earth Planet. Sci. Lett., 84: 471-478.

Forsyth, D.W., 1991. Comment on "A quantitative study of the axial topography of the Mid-Atlantic Ridge" by A. Malinverno. J. Geophys. Res., 96: 2039-2047.

Francheteau, J., Armijo, R., Cheminee, J.L., Hékinian, R., Lonsdale, P. and Blum, N., 1990. 1 Ma East Pacific Rise oceanic crust and uppermost mantle exposed by rifting in Hess Deep (Equatorial Pacific Ocean). Earth Planet. Sci. Lett., 101: 281-295.

Géli, L., 1993. Volcano-tectonic events and sedimentation since Late Miocene time at the Mohns Ridge, near 72°N, in the Norwegian-Greenland Sea. Tectonophysics, 222: 417-444.

Géli, L., Ondreas, H., Olivet, J.-L., Sahabi, M., Aslanian, D. and Gilg Capar, L., 1994a. Thermal structure vs. spreading rate at intermediate spreading rates: the example of the Pacific-Antarctic Ridge between

55°S and 63°S. Eos, Trans. AGU, 7: 330.
Géli, L., Renard, V. and Rommevaux, C., 1994b. Ocean crust formation processes at very slow spreading centers: a model for the Mohns ridge, near 72°N, based on magnetic, gravity, and seismic data. J. Geophys. Res., 99: 2995-3013.
Gente, P., 1987. Etude morphostructurale comparative de dorsales océaniques à taux d'expansion variés., Thesis, Université de Bretagne Occidentale (Brest), 371 pp.
Gente, P., Pockalny, R.A., Durand, C., Deplus, C., Maia, M., Ceuleneer, G., Mével, C., Cannat, M. and Laverne, C., (in press), Characteristics and evolution of the segmentation of the Mid-Atlantic Ridge between 20°N and 24°N during the last 10 million years. Earth Planet. Sci. Lett. in press.
Girardeau, J. and Francheteau, J., 1993. Plagioclase-wehrlites and peridotites on the East Pacific Rise (Hess Deep) and the Mid-Atlantic Ridge (DSDP site 334): evidence for magma percolation in the oceanic upper mantle. Earth Planet. Sci. Lett., 115: 137-149.
Gracia, E., Ondreas, H., Bendel, V. and the STARMER Group, 1994. Multi-scale morphologic variability of the North Fiji Basin Ridge (Southwest Pacific). Mar. Geol., 116: 133-151.
Grindlay, N.R., Fox, P.J. and Vogt, P.R., 1992. Morphology and tectonics of the Mid-Atlantic Ridge (25°-27°30'S) from Sea Beam and magnetic data. J. Geophys. Res., 97: 6983-7010.
Hamlyn, P.R. and Bonatti, E., 1980. Petrology of mantle-derived ultramafics from the Owen Fracture Zone, Northwest Indian Ocean: implications for the nature of the oceanic upper mantle. Earth Planet. Sci. Lett., 48: 65-79.
Harper, G.D., 1985. Tectonics of slow-spreading mid-ocean ridges and consequences of a variable depth to the brittle/ductile transition. Tectonics, 4: 395-409.
Hébert, R., Bideau, D. and Hékinian, R., 1983. Ultramafic and mafic rocks from the Garret transform fault near 13°30'S on the East Pacific Rise: igneous petrology. Earth Planet. Sci. Lett., 65: 107-125.
Hékinian, R. and Aumento, F., 1973. Rocks from the Gibbs fracture zone and the Minia seamount near 53°N in the Atlantic Ocean. Mar. Geol., 14: 47-72.
Hékinian, R., Bideau, D., Cannat, M., Francheteau, J. and Hébert, R., 1992. Volcanic activity and crust-mantle exposure in the ultrafast Garrett Transform Fault near 13°28'S in the Pacific. Earth Planet. Sci. Lett., 108: 259-275.
Hékinian, R., Bideau, D., Francheteau, J., Cheminee, J.-L., Armijo, R., Londsale, P. and Blum, N., 1993. Petrology of the East Pacific Rise crust and upper mantle exposed in the Hess Deep (Eastern Equatorial Pacific). J. Geophys. Res., 98: 8069-8094.
Hodges, F.N. and Papike, J.J., 1977. Petrology of basalts, gabbros, and peridotites from DSDP leg 37. Initial Rep. Deep Sea Drill. Proj., 37: 711-719.
Honnorez, J. and Kirst, P., 1975. Petrology of rodingites from the Equatorial Mid-Atlantic Ridge fracture zones and their geotectonic significance. Contrib. Mineral. Petrol., 49: 233-257.
Huang, P.Y., Solomon, S.C., Bergman, E.A. and Nabelek, J.L., 1986. Focal depths and mechanisms of Mid-Atlantic Ridge earthquakes from body waveform inversion. J. Geophys. Res., 91: 579-598.
Huang, P.Y. and Solomon, S.C., 1988. Centroid depths of Mid-Ocean-Ridge earthquakes: dependence on spreading rate. J. Geophys. Res., 93: 13445-13477.
Ito, E. and Anderson, A.T., 1983. Submarine metamorphism of gabbros from the Mid-Cayman Rise: petrographic and mineralogic contraints on hydrothermal processes at slow-spreading ridges. Contrib. Mineral. Petrol., 82: 371-388.
Jaques, A.L. and Green, D.H., 1980. Anhydrous melting of peridotite at 0-15 kb pressure and the genesis of tholeiitic basalts. Contrib. Mineral. Petrol., 73: 287-310.
Johnson, K.T.M. and Dick, H.J.B., 1992. Open system melting and temporal and spatial variation of peridotite and basalt at the Atlantis II Fracture Zone. J. Geophys. Res., 97: 9219-9241.
Juteau, T., Berger, E. and Cannat, M., 1990a. Serpentinized, residual mantle peridotites from the MAR median valley, ODP hole 670a (21°10'N, 45°02'W, leg 109): primarily mineralogy and geothermometry. Proc. Ocean Drill. Prog., Init. Repts., 106/109: 27-45.
Juteau, T., Cannat, M. and Lagabrielle, Y., 1990b. Serpentinized peridotites in the upper oceanic crust away from transform zones: a comparison of the results of previous DSDP and ODP legs. Proc. Ocean Drill. Prog., Init. Repts., 106/109: 303-308.
Kappel, E.S. and Ryan, W.B.F., 1986. Volcanic episodicity and a non-steady state rift valley along northeast Pacific spreading centers: evidence from Sea MARC I. J. Geophys. Res., 91: 13925-13940.
Karson, J.A., Brown, J.R. and Winters, A.T., 1986. Seafloor spreading in the MARK area. Eos, Trans. AGU, 67: 1213.
Karson, J.A., Thompson, G., Humphris, S.E., Edmond, J.M., Bryan, W.B., Brown, J.R., Winters, A.T., Pockalny, R.A., Casey, J.F., Campbell, A.C., Klinkhammer, G., Palmer, M.R., Kinzler, R.J. and Sulanowska, M.M., 1987. Along-axis variations in seafloor spreading in the MARK area. Nature, 328: 681-685.

Karson, J.A., 1991. Seafloor spreading on the Mid-Atlantic Ridge: implications for the structure of ophiolites and oceanic lithosphere produced in slow-spreading environments. In: J. Malpas, E.M. Moores, A. Panayiotou and C. Xenophontos (Eds), Proceedings of the Symposium "Troodos 1987". Geological Survey Department, Nicosia, Cyprus, pp. 547-555.

Karsten, J.L., Hammond, S.R., Davis, E.E. and Currie, R.G., 1986. Detailed geomorphology and neotectonics of the Endeavour Segment, Juan de Fuca Ridge: new results from Seabeam swath mapping. Geol. Soc. Am. Bull., 97: 213-221.

Kashintsev, G.L., Kuzmin, M.L. and Popolitov, E.N., 1982. Composition and structure of the oceanic crust in the vicinity of the Hess Basin (Pacific Ocean). Geotectonics, 16: 512-520.

Klein, E.M. and Langmuir, C.H., 1987. Global correlations of ocean ridge basalt chemistry with axial depth and crustal thickness. J. Geophys. Res., 92: 8089-8115.

Klein, E.M., Karsten, J.L., Batiza, R., Bailey, J., Bordelon, M., Broda, J., Ferguson, E., Gowen, M., Mukhopadhyay, R., Pichler, T., Sherman, S., Stephani, R., Tassara, A., Thatcher, M. and Young, R., 1993. Results from the 1993 southern CROSS expedition: morphological variations along the southern Chile Ridge. Eos, Trans. AGU, 74: 686.

Kong, L.S.L., Solomon, S.C. and Purdy, G.M., 1992. Microearthquake characteristics of a mid-ocean ridge along-axis high. J. Geophys. Res., 97: 1659-1685.

Kuo, B.-Y. and Forsyth, D.W., 1988. Gravity anomalies of the ridge-transform system in the South Atlantic between 31 and 34.5°S: upwelling centers and variations in the crustal thickness. Mar. Geophys. Res., 10: 205-232.

Lafoy, Y., Auzende, J.-M., Ruellan, E., Huchon, P. and Honza, E., 1990. The 16°40'S triple junction in the North Fiji Basin (SW Pacific). Mar. Geophys. Res., 12: 285-296.

Lagabrielle, Y. and Cannat, M., 1990. Alpine jurassic ophiolites resemble the modern central Atlantic basement. Geology, 18: 319-322.

Lagabrielle, Y., Mamaloukas-Frangoulis, V., Cannat, M., Auzende, J.-M., Honnorez, J., Mével, C. and Bonatti, E., 1992. Vema Fracture Zone (Central Atlantic): tectonic and magmatic evolution of the median ridge and the eastern ridge-transform intersection domain. J. Geophys. Res., 97: 17331-17351.

Langmuir, C.H., Bender, J.F., Shirey, S. and the Venture Leg 2 Scientific Team, 1990. A trace-element enriched province on the East Pacific Rise north of the Orozco Transform Fault. Eos, Trans. AGU, 71: 1703.

Lewis, B.T.R., 1979. Periodicities in volcanism and longitudinal magma flow on the East Pacific Rise at 23°N. Geophys. Res. Lett., 6: 753-756.

Lichtman, G.S. and Eissen, J.-P., 1983. Time and space contraints on the evolution of medium-rate spreading centers. Geology, 11: 592-595.

Lin, J., Purdy, G.M., Schouten, H., Sempere, J.-C. and Zervas, C., 1990. Evidence from gravity data for focussed magmatic accretion along the Mid-Atlantic Ridge. Nature, 344: 627-632.

Lin, J. and Phipps Morgan, J., 1992. The spreading rate dependence of three-dimensional mid-ocean ridge gravity structure. Geophys. Res. Lett., 19: 13-16.

Lonsdale, P., 1986. Tectonic and magmatic ridges in the Eltanin fault system, South Pacific. Mar. Geophys. Res., 8: 203-242.

Macdonald, K.C., 1982. Mid-ocean ridges: fine scale tectonic, volcanic and hydrothermal processes within the plate bounbary zone. Ann. Rev. Earth Planet. Sci., 10: 155-190.

Madsen, J.A., Detrick, R.S., Mutter, J.C., Buhl, P. and Orcutt, J.A., 1990. A two- and three-dimensional analysis of gravity anomalies associated with the East Pacific Rise at 9°N and 13°N. J. Geophys. Res., 95: 4967-4987.

Malcolm, F.L., 1981. Microstructures of the Cayman Trough gabbros. J. Geol., 89: 675-688.

Malinverno, A. and Pockalny, R.A., 1990. Abyssal hill topography as an indicator of episodicity in the crustal accretion and deformation. Earth Planet. Sci. Lett., 99: 154-169.

Malinverno, A., 1991. Inverse square-root dependence of mid-ocean-ridge flank roughness on spreading rate. Nature, 352: 58-60.

Marks, K.M. and Stock, J.M., 1994. Variations in ridge morphology and depth-age relationships on the Pacific-Antarctic Ridge. J. Geophys. Res., 99: 531-541.

McAdoo, D.C. and Marks, K.M., 1992. Gravity fields of the Southern Ocean from Geosat data. J. Geophys. Res., 97: 3247-3260.

McClain, J.S. and Lewis, B.T.R., 1980. A seismic experiment at the axis of the East Pacific Rise. Mar. Geol., 35: 147-169.

McKenzie, D., 1967. Some remarks on heat flow and gravity anomalies. J. Geophys. Res., 72: 6261-6273.

McKenzie, D., 1984. The generation and compaction of partially molten rock. J. Petro., 25: 713-765.

Melson, W.G. and Thompson, G., 1970. Layered basic complex in oceanic crust, Romanche Fracture, Equatorial Atlantic Ocean. Science, 168: 817-820.

Menard, H.W., 1967. Sea floor spreading, topography, and the second layer. Science, 157: 923-924.

Mével, C., Auzende, J.-M., Cannat, M., Dorval, J.-P., Dubois, J., Fouquet, Y., Gente, P., Grimaud, P., Karson, J.A., Segonzac, M. and Stievenard, M., 1988. HYDROSNAKE 1988: submersible study of seafloor spreading in the MARK area. Eos, Trans. AGU, 69: 1439-1440.

Mével, C., Cannat, M., Gente, P., Marion, E., Auzende, J.-M. and Karson, J.A., 1991. Emplacement of deep crustal and mantle rocks on the west median valley wall of the MARK area (MAR, 23°N). Tectonophysics, 190: 31-53.

Michael, P.J. and Bonatti, E., 1985a. Peridotite composition from the North Atlantic: regional and tectonic variations and implications for partial melting. Earth Planet. Sci. Lett., 73: 91-104.

Michael, P.J. and Bonatti, E., 1985b. Petrology of ultramafic rocks from sites 556, 558, and 560 in the north Atlantic. Initial Rep. Deep Sea Drill. Proj., 82: 523-530.

Miyashiro, A., Shido, F. and Ewing, M., 1969. Composition and origin of serpentinites from the Mid-Atlantic Ridge near 24° and 30° north latitude. Contrib. Mineral. Petrol., 23: 117-127.

Morris, E. and Detrick, R.S., 1991. Three-dimensional analysis of gravity anomalies in the MARK area, Mid-Atlantic Ridge 23°N. J. Geophys. Res., 96: 4355-4366.

Morton, J.L. and Sleep, N.H., 1985. A mid-ocean ridge thermal model: contraints on the volume of axial hydrothermal heat flux. J. Geophys. Res., 90: 11345-11353.

Murray, M.H., Kong, L., Forsyth, D.W., Solomon, S.C. and Hussong, D.M., 1984. An OBS study of microearthquakes in the median valley on the Mid-Atlantic Ridge near 35°N. Eos, Trans. AGU, 65: 1009.

Neumann, G.A. and Forsyth, D.W., 1993. The paradox of the axial profile: isostatic compensation along the axis of the Mid-Atlantic Ridge? J. Geophys. Res., 98: 17891-17910.

Nicolas, A., 1989. Structures of ophiolites and dynamics of oceanic lithosphere., Kluwer Academic Publishers, 368 pp.

Nicolas, A., Boudier, F. and Bouchez, J.-L., 1980. Interpretation of peridotite structures from ophiolitic and oceanic environments. Am. J. Sci., 280: 192-280.

Niu, Y. and Batiza, R., 1993. Chemical variation trends at fast and slow spreading Mid-Ocean Ridges. J. Geophys. Res., 98: 7887-7902.

OTTER, 1984. The geology of the Oceanographer transform: the ridge-transform intersection. Mar. Geophys. Res., 6: 109-141.

Palmer, J., Sempere, J.-C., Christie, D.M. and Phipps Morgan, J., 1993. Morphology and tectonics of the Australian-Antarctic Discordance between 123°E and 128°E. Mar. Geophys. Res., 15: 121-152.

Parmentier, E.M. and Phipps Morgan, J., 1990. Spreading rate dependence of three-dimensional structure in oceanic centres. Nature, 348: 325-328.

Parsons, B.E. and Sclater, J.G., 1977. An analysis of the variation of ocean floor bathymetry with age. J. Geophys. Res., 82: 803-827.

Patriat, P. and Courtillot, V., 1984. On the stability of triple junctions and its relation to episodicity in spreading. Tectonics, 3: 317-332.

Peaceman, D.W. and Rachford, H.H.J., 1955. Numerical solution of parabolic and elliptic differential equations. J. Soc. Indus. Appl. Math., 3: 28-41.

Phillips, J.D., Thompson, G., Von Herzen, R.P. and Bowen, V.T., 1969. Mid-Atlantic Ridge near 43°N latitude. J. Geophys. Res., 74: 3069-3081.

Phipps Morgan, J. and Forsyth, D.W., 1988. Three-dimensional flow and temperature perturbations due to a transform offset: effects on oceanic crustal and upper mantle structure. J. Geophys. Res., 93: 2955-2966.

Pockalny, R.A., Detrick, R.S. and Fox, P.J., 1988. Morphology and tectonics of the Kane Transform from Sea Beam bathymetry data. J. Geophys. Res., 93: 3179-3193.

Prinz, M., Keil, K., Green, J.A., Reid, A.M., Bonatti, E. and Honnorez, J., 1976. Ultramafic and mafic dredge samples from the Equatorial Mid-Atlantic Ridge and fracture zones. J. Geophys. Res., 81: 4087-4103.

Riedesel, M., Orcutt, J.A., Macdonald, K.C. and McClain, J.S., 1982. Microearthquakes in the black smoker hydrothermal field, East Pacific Rise at 21°N. J. Geophys. Res., 87: 10613-10623.

Rommevaux, C., 1994. Etude gravimétrique et magnétique de l'évolution de la segmentation des dorsales lentes., Thesis, Université de Paris VII, 282 pp.

Rommevaux, C., Deplus, C., Patriat, P. and Sempéré, J.-C., 1994. Three-dimensional gravity study of the Mid-Atlantic Ridge: evolution of the segmentation between 28° and 29°N during the last 10 m.y. J. Geophys. Res., 99: 3015-3029.

Rona, P.A., Boström, K., Laubier, L. and Smith, K.L., 1983. Hydrothermal processes at seafloor spreading centers., Plenum press, 796 pp.

Rona, P.A., Widenfalk, L. and Boström, K., 1987. Serpentinized ultramafics and hydrothermal activity at

the Mid-Atlantic Ridge crest near 15°N. J. Geophys. Res., 92: 1417-1427.
Rudnik, G.B., 1976. Magmatic and metamorphism rocks in Hess Deep (in russian)., Geological and geophysical researches in the southeastern part of the Pacific Ocean. Nauka, Moscow, pp. 116-125.
Ruegg, J.-C., Kasser, M. and Lépine, J.-C., 1984. Strain accumulation across the Asal-Ghoubbet Rift, Djibouti, East Africa. J. Geophys. Res., 89: 6237-6246.
Sandwell, D.T., 1992. Antarctic marine gravity field from high-density satellite altimetry. Geophys. J. Int., 109: 437-448.
Sauter, D., Whitechurch, H., Munschy, M. and Humler, E., 1991. Periodicity in the accretion process on the Southeast Indian Ridge at 27°40'S. Tectonophysics, 195: 47-64.
Schulz, N.J., Detrick, R.S. and Miller, S.P., 1988. Two and three dimensional inversions of magnetic anomalies in the MARK area (Mid-Atlantic Ridge 23°N). Mar. Geophys. Res., 10: 41-57.
Sempéré, J.-C., Lin, J., Brown, H.S., Schouten, H. and Purdy, G.M., 1993. Segmentation and morphotectonic variations along a slow-spreading center: the Mid-Atlantic Ridge (24°00'N-30°40'N). Mar. Geophys. Res., 15: 153-200.
Shipboard Scientific Party, Aumento, F., Melson, W.G., Bougault, H., Dmitriev, L., Fisher, J.F., Flower, M., Hall, J.M., Howe, R.C., Hyndman, R.D., Miles, G.A. and Robinson, P.T., 1977. Site 334. Initial Rep. Deep Sea Drill. Proj., 37: 239-287.
Shipboard Scientific Party, Bryan, W.B., Juteau, T., Adamson, A.C., Autio, L.K., Becker, K., Bina, M., Eissen, J.-P., Fijii, T., Grove, T.L., Hamano, Y., Hebert, R., Komor, S.C., Kopietz, J., Krammer, K., Loubet, M., Moos, D. and Richards, H.G., 1988. Site 670. Proc. Ocean Drill. Prog., Init. Repts., 109: 203-237.
Sinton, J.M., 1979. Petrology of (alpine-type) peridotites from site 395, DSDP leg 45. Initial Rep. Deep Sea Drill. Proj., 45: 595-601.
Sleep, N.H., 1969. Sensitivity of heat flow and gravity to the mechanism of sea-floor spreading. J. Geophys. Res., 74: 542-549.
Sleep, N.H., 1974. Segregation of magma from a mostly crystalline mush. Geol. Soc. Am. Bull., 85: 1225-1232.
Sleep, N.H., 1975. Formation of oceanic crust: some thermal contraints. J. Geophys. Res., 80: 4037-4042.
Sloan, H. and Patriat, P., 1992. Kinematics of the North American-African plate boundary between 28° and 29°N during the last 10 Ma: evolution of the axial geometry and spreading rate and direction. Earth Planet. Sci. Lett., 113: 323-341.
Small, C. and Sandwell, D.T., 1992. An analysis of ridge axis gravity roughness and spreading rate. J. Geophys. Res., 97: 3235-3245.
Sotin, C. and Parmentier, E.M., 1989. Dynamical consequences of compositional and thermal density stratification beneath spreading centers. Geophys. Res. Lett., 16: 835-838.
Spiegelman, M. and McKenzie, D., 1987. Simple 2-D models for melt extraction at mid-ocean ridges and island arcs. Earth Planet. Sci. Lett., 83: 137-152.
Stein, R.S., Briole, P., Ruegg, J.-C., Tapponnier, P. and Gasse, F., 1991. Comtemporary, holocene, and quaternary deformation of the Asal Rift, Djibouti: implications for the mechanics of slow spreading ridges. J. Geophys. Res., 96: 21789-21806.
Stroup, J.B. and Fox, P.J., 1981. Geologic investigations in the Cayman Trough: evidence for thin oceanic crust along the Mid-Cayman Rise. J. Geol., 89: 395-420.
Tapponnier, P. and Francheteau, J., 1978. Necking of the lithosphere and mechanics of slowly accreting plate boundaries. J. Geophys. Res., 83: 3955-3970.
Thompson, G. and Melson, W.G., 1972. The petrology of oceanic crust across fracture zones in the Atlantic Ocean: evidence of a new kind of seafloor spreading. J. Geol., 80: 526-538.
Tiezzi, L.J. and Scott, R.B., 1980. Crystal fractionation in a cumulate gabbro, Mid-Atlantic Ridge, 26°N. J. Geophys. Res., 85: 5438-5454.
Tisseau, C. and Tonnerre, T., In preparation. Non steady-state thermal model of slow spreading centers: role of cooling periods and implications for the axial domain.
Tolstoy, M., Harding, A.J. and Orcutt, J.A., 1993. Crustal thickness on the Mid-Atlantic Ridge: bull's eye gravity anomalies and focused accretion. Science, 262: 726-729.
Toomey, D.R., Solomon, S.C., Purdy, G.M. and Murray, M.H., 1985. Microearthquakes beneath the median valley of the Mid-Atlantic Ridge near 23°N: hypocenters and focal mechanisms. Geophys. Res. Lett., 90: 5443-5458.
Toomey, D.R., Solomon, S.C. and Purdy, G.M., 1988. Microearthquakes beneath the median valley of the Mid-Atlantic Ridge near 23°N: tomography and tectonics. J. Geophys. Res., 93: 9093-9112.
Toomey, D.R., Purdy, G.M., Solomon, S.C. and Wilcock, W.S.D., 1990. The three-dimensional seismic velocity structure of the East Pacific Rise near latitude 9°30'N. Nature, 347: 639-645.
Tréhu, A.M. and Solomon, S.C., 1983. Earthquakes in the Orozco Transform Zone: seismicity, source

mechanisms, and tectonics. J. Geophys. Res., 88: 8203-8225.
Vanko, D.A. and Batiza, R., 1982. Gabbroic rocks from the Mathematician Ridge failed rift. Nature, 300: 742-744.
Vaslet, N., 1993. Apports des images du sonar latéral SAR à l'étude de la structure fine des dorsales rapides. Implications sur les relations entre tectonique - magmatisme - hydrothermalisme (exemple de la Dorsale Est-Pacifique entre 13°20'N et 11°50'N)., Thesis, Université de Bretagne Occidentale (Brest), 336 pp.
Wakita, H., Nagasawa, H., Uyeda, S. and Kuno, H., 1967. Uranium, thorium and potassium contents of possible mantle materials. Geochem. J., 1: 183.
Weissel, J.K. and Hayes, D.E., 1974. The Australian-Antartic Discordance: new results and implications. J. Geophys. Res., 79: 2579-2587.
Whitehead, J.A., Dick, H.J.B. and Schouten, H., 1984. A mechanism for magmatic accretion under spreading centres. Nature, 312: 146-148.
Wilcock, W.S.D., Purdy, G.M., Solomon, S.C., DuBois, D.L. and Toomey, D.R., 1992. Microearthquakes on and near the East Pacific Rise, 9°-10°N. Geophys. Res. Lett., 19: 2131-2134.
Wolfe, C.J., Purdy, G.M., Toomey, D.R. and Solomon, S.C., 1993. FARA microearthquake experiments II: results from the Mid-Atlantic Ridge at 29°N. Eos, Trans. AGU, 74: 601.

Petrology and Structural Geology

1. J.P. Bard: *Microtextures of Igneous and Metamorphic Rocks.* 1986
ISBN Hb: 90-277-2220-X; ISBN Pb: 90-277-2313-3
2. A. Nicolas: *Principles of Rock Deformation.* 1987.
ISBN Hb: 90-277-2368-0; ISBN Pb: 90-277-2369-9
3. J.D. Macdougall (ed.): *Continental Flood Basalts.* 1988.
ISBN 90-277-2806-2
4. A. Nicolas: *Structures of Ophiolites and Dynamics of Oceanic Lithosphere.* 1989
ISBN 0-7923-0255-9
5. Tj. Peters, A. Nicolas and R.G. Coleman (eds.): *Ophiolite Genesis and Evolution of the Oceanic Lithosphere.* Proceedings of the Ophiolite Conference Muscat, Oman, January 1990). 1991
ISBN 0-7923-1176-0
6. R.L.M. Vissers and A. Nicolas (eds.): *Mantle and Lower Crust Exposed in Oceanic Ridges and in Ophiolites.* Contributions to a Specialized Symposium of the VII EUG Meeting, Strasbourg, Spring 1993. 1995
ISBN 0-7923-3491-4

KLUWER ACADEMIC PUBLISHERS – DORDRECHT / BOSTON / LONDON

GPSR Compliance

The European Union's (EU) General Product Safety Regulation (GPSR) is a set of rules that requires consumer products to be safe and our obligations to ensure this.

If you have any concerns about our products, you can contact us on ProductSafety@springernature.com

In case Publisher is established outside the EU, the EU authorized representative is:

Springer Nature Customer Service Center GmbH
Europaplatz 3
69115 Heidelberg, Germany

Batch number: 10583998

Printed by Printforce, the Netherlands